The Geological Sciences in the Antebellum South

THE GEOLOGICAL SCIENCES IN THE ANTEBELLUM SOUTH

EDITED BY

James X. Corgan

THE UNIVERSITY OF ALABAMA PRESS
Tuscaloosa

The University of Alabama Press
Tuscaloosa, Alabama 35487-0380
uapress.ua.edu

Hardcover edition published 1982.
Paperback edition published 2014.
eBook edition published 2014.

Manufactured in the United States of America

Grateful acknowledgment is made to Austin Peay State University, whose Tower Fund provided a faculty research grant for manuscript expenses.

∞

The paper on which this book is printed meets the minimum requirements of American National Standard for Information Science–Permanence of Paper for Printed Library Materials, ANSI Z39.48-1984.

Paperback ISBN: 978-0-8173-5798-6
eBook ISBN: 978-0-8173-8793-8

A previous edition of this book has been catalogued by the Library of Congress as follows:

Main entry under title:
The Geological sciences in the antebellum South.
Papers originally presented at a symposium held during a Southeastern sectional meeting of the Geological Society of America in Birmingham, March 1980.
Bibliography: p.
Includes index.
Contents: Introduction — Andrew Ellicott's geological observations in the Mississippi Valley and Florida in 1796–1800 / by George W. White — South Carolina state geological surveys of the nineteenth century / by Anne Millbrooke — Early American geological surveys and Gerard Troost's field assistants / by James X. Corgan — [etc.]
1. Geology—Southern States—History—Congresses. I. Corgan, James X. II. Geological Society of America. Southeastern Section.
QE13.U6G46 557.5'09 81-2993
ISBN 0-8173-0076-7 AACR2

Contents

The Geological Sciences in the Antebellum South

Introduction

Some years ago C. P. Snow caught the public fancy with a lecture on the twin cultures of modern time, the scientific and the nonscientific.[1] The historical implications of Snow's analysis are disturbing. People of the nonscientific culture normally keep the records of civilization, for this group includes the historians, who control what posterity learns. Typically, historians study just part of civilization and they emphasize the part they find most interesting: the nonscientific culture in which they are personally involved. The things of science rarely interest the nonscientist and beyond mere lack of interest there can be blatant cultural prejudice: a deemphasis of things that do not arouse feelings of empathy.

In studies of southern history, the individual scientific fields and science as a whole seem to suffer from systematic oversight. At a minimum, they have not been overstudied. Yet science is deeply integrated into contemporary southern culture. Today, birdwatchers crawl all over the southern landscape, spelunkers penetrate the underground, every county has soil conservationists, modern medical practitioners belong to local technical societies, and the government of every southern state employs a growing multitude of professional scientists. Each of these scientific groups has a history that could be studied. Very few have been studied in detail, and even fewer have been studied critically.

In our complex society it may seem unfair to charge one group, the historians, with the general task of studying all aspects of the past. It is unfair, but it is necessary. Lord Snow is right; our culture is growing schismatic, and therefore we have a great need to generalize and to synthesize. There has always been a need for synthesis. People understood this millenia ago, when they named just nine Muses for all the ancient liberal disciplines. Poor, harried Clio is the only Muse for all of the details and all of the generalizations of history. In striving for a better understanding of southern history, one of Clio's greatest problems is a lack of data on the scientific side of the general culture.

For the last decade, I have been deeply involved in studies that examine the evolution of the sciences in general, and geology in particular, within the culture of the southern states. From personal experience I know there is a need for broad, interdisciplinary studies that place each aspect of southern science within a national perspective. These studies must also bridge the widening gap between the twin cultures of Lord Snow. In 1977 I therefore began to search for scholars who might participate in a general symposium focused on one area of science, the geological sciences, as it developed in the South prior to 1861.

In March 1980 I chaired a symposium entitled "The History of the Geological Sciences in the Antebellum South," held during a southeastern sectional meeting of the Geological Society of America in Birmingham. Nine papers were part of that symposium.[2] They touched on the intellectual history of twenty states (Figure 1) and are presented here in edited form. The text that follows this introduction begins with a paper focused on observations made from 1796 through 1800 and ends with a study of work done during the 1850s. Following the text are brief biographical sketches of the contributors, a bibliography, and an index.

The scientific and the nonscientific cultures are about equally represented by the ten contributors who have made this symposium possible. As their writings suggest, all contributors have some status as students of history, but in terms of primary background there are four geologists, one chemist, one zoologist, one librarian, and three historians. Each contributor comes from a different state and only three are southerners. All contributors are experienced scholars who blend science and nonscience with sufficient intimacy to work comfortably at the interface between the twin cultures.

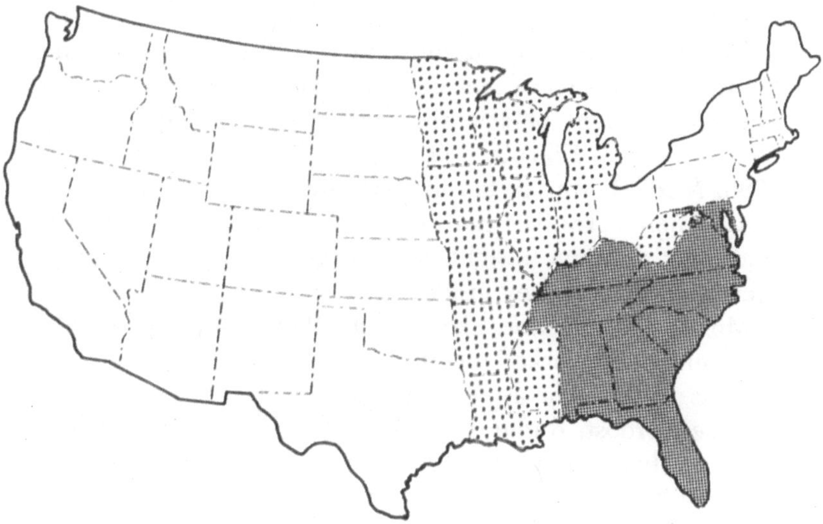

Figure 1.

AREA STUDIED. Symposium contributions provide detailed information on the early geological exploration of states shown in the dark pattern. States shown in the light pattern are discussed in less detail. For a summary of the geographic aspects of this symposium see state names in the index at the end of the volume.

Most contributions are at least partially biographical, studying the careers of people who worked in geology and allied fields. The following scientists are discussed at reasonable length:

Richard Owen Currey	Joseph Nicollet
Andrew Ellicott	David Dale Owen
Ebenezer Emmons	William Barton Rogers

Oscar Lieber
Abram Litton
Charles Lyell
Edmund Ruffin
Gerard Troost
Michael Tuomey
Lardner Vanuxem

Even to specialists in southern intellectual history, many names in this list may be unfamiliar. Yet all worked in the South and all had a significant impact on the development of scientific thought or scientific institutions. Three pioneers were travelers: Ellicott, who began his southern travels in 1796; Nicollet, who began in 1831; and Lyell, who began in 1841. Their narratives not only provide geological observations, but also give insight into the general intellectual and social life of areas they visited.

Most of these pioneers knew each other, at least by reputation, and many worked together. Several made major contributions to different aspects of the growth of the geological sciences. Troost, Ruffin, Owen, and Nicollet are discussed at length in two of the essays in this volume. Nicollet read the works of Ellicott and made similar observations; each pioneer who was active in the 1830s or later probably knew of Ruffin's research. There are many other evidences of interrelationships.

While several essays are primarily biographical, others are not. One treats the development of a specific application of the earth sciences, the use of mineral fertilizers, and others discuss the development of geological surveys in three different states: South Carolina, Virginia, and Kentucky. These studies do not emphasize biography, yet each blends biography with the history of ideas and the history of institutions. Similarly, most contributions that are explicitly biographical also provide data on the history of ideas and the history of institutions that employed geologists. Thus each study contains the same basic elements.

These essays examine the years from 1796 to 1860. During that time the nature of science and of geology in particular changed. In 1796 the word *geology* was not universally used for the academic discipline we call geology today, and the discipline itself was much less clearly defined. To cite just one example, the scientific study of the earth, which we now call geology, in the early days often included much of the scientific study of soils, which we now call pedology and regard as a separate science.

There are even more fundamental differences between modern concepts and the concepts of the pioneers. Today, geologist is a specialized profession that falls under the catchall term, *scientist*. Every living American has some idea of the role the scientist plays in contemporary society. The word *scientist* was coined in 1834.[3] In the early years covered by this symposium no one earned a livelihood through geology or through science; there was no such thing as a scientist. By the end of the antebellum era, geology and several other sciences were well-established professional fields. To some rather modest extent, this symposium contributes to a general recording of the growth of scientific institutions and the professionalization of science in the South.

Finally, this symposium casts some light on an epic adventure in the history of human thought; the development of the concept that the planet Earth has an orderly history that can be studied scientifically. Some current thoughts on the planet's past are shown in Figure 2. Earth is now believed to be well over 3.8 billion years old, and the last 570 million years are subdivided into named units that can be recognized all over the globe.[4]

The worldwide correlation of time units was a concern of many geologists during the years under discussion. The essays that deal with Charles Lyell and with Ebenezer Emmons delve into intercontinental correlations. Authors of these and other contributions occasionally use a standard term from the geological time scale, but they are careful to avoid unnecessary detail.

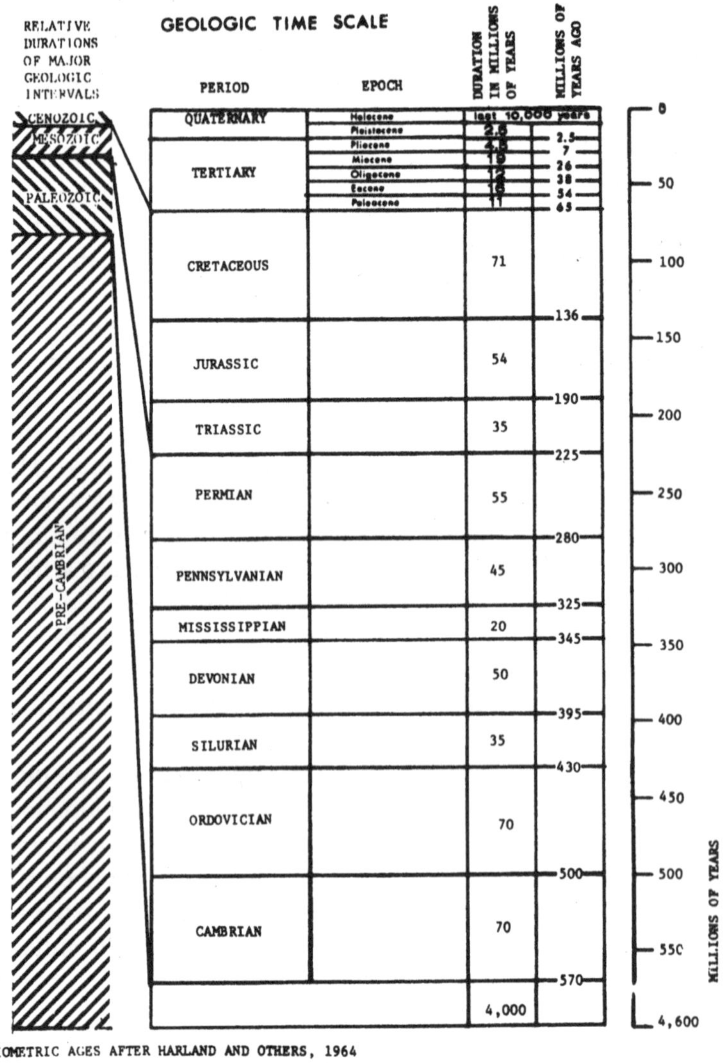

Figure 2.

THE GEOLOGICAL COLUMN. During the years covered by this symposium, concepts of the history of Earth were subjects of heated discussion. Several southern scientists became involved in the development of concepts of earth history. Terms used in this illustration represent current preferences and are occasionally used in the symposium papers. This illustration is, essentially, a glossary for time terms used in the symposium contributions.

There is an active avoidance of detail in all the essays in this volume. Nicollet's southern travels are reviewed without going into the details of how his barometers were calibrated. Similarly, Troost's influence on field work is appraised without identifying the distinctive technique he used to estimate coal resources. As anyone who writes on any historical topic soon learns, there are a variety of audiences toward which a study can be directed; all contributions to the present symposium are directed toward the most general audience. No contribution assumes a background in any of the sciences or an especially strong interest in the scientific aspects of modern culture.

Those who arrange and edit papers presented at symposia can search too hard for similarities and for common threads. The great strength of symposia is their heterogeneity, especially the broad range of expertise that can be focused on a subject. The symposium that gave rise to this volume involved nine very different papers. They fit together, but they do not blend.

The value of a published symposium is influenced by writing and by editing, but depends mostly upon the topic studied. Within the culture of the southern states, science as a whole, the individual sciences, and scientific institutions are far from favorite subjects for historical study. The growth of scientific aspects of southern culture is chiefly known from ancient tomes that verge on boosterism, like Johnson's *Scientific Interests in the Old South*, and from nonscience-oriented intellectual histories, like Eaton's *The Mind of the Old South*.[5] A dearth of contemporary scholarship should enhance the value of the present volume.

James X. Corgan

Notes

1. C. P. Snow, *The Two Cultures* and *The Scientific Revolution* (Cambridge: At the University Press, 1959).

2. Abstracts of all contributions were published in volume 12 (1980) of the *Geological Society of America, Abstracts with Programs:* Michele L. Aldrich, "William Barton Rogers and the Virginia Geological Survey 1835-1842," p. 169. Daniel D. Arden, "Charles Lyell's Observation on Southeastern Geology," pp. 169–70. Martha C. Bray, "Southern Influences on the Career of Joseph Nicollet," p. 172. James X. Corgan, "Troost's Field Assistants, 1831–1836," p. 175. Markes E. Johnson, "The Second Geological Career of Ebenezer Emmons: Success and Failure in the Southern States (1851–1860)," p. 180. Anne Millbrooke, "South Carolina Geological Surveys of the Nineteenth Century," p. 201. Richard C. Sheridan, "Mineral Fertilizers in Southern Agriculture," p. 208. George W. White, "Andrew Ellicott's Geological Observations in the Mississippi Valley and Florida in 1802–1804," p. 212. Ivan L. Zabilka, "Antebellum Geological Surveys in Kentucky and their Contribution to the Shaler Surveys of the 1870's," p. 213.

3. For a documented discussion of the words *geology* and *scientist*, see the essay on Gerard Troost.

4. W. B. Harland, et al., *The Phanerozoic Time-Scale* (London: The Geological Society of London, 1964).

5. Thomas Cary Johnson, Jr., *Scientific Interests in the Old South* (New York: Appleton-Century, 1936); Clement Eaton, *The Mind of the Old South*, rev. ed. (Baton Rouge: Louisiana State University Press, 1964).

Andrew Ellicott's Geological Observations in the Mississippi Valley and Florida, 1796–1800

George W. White

Almost all of the data on the history of early American geology must be sought in reports of travels such as Andrew Ellicott's. The first book on the geology of the United States, written by J. D. Schoepf in 1787, was in German and remained almost unknown. Similarly, a French-language report by C. F. Volney contained a great deal of geological information, but remained little known to Americans, although translated into English in 1804. The first widely read work on the geology of the United States was an English-language account by William Maclure that appeared in 1809.[1]

In the years before Maclure's report, travel books were a major medium for communicating geological observations. A few early travel writers collected their scientific observations into special sections of their books, but most did not. After 1750 some writers speculated on the origin of the geological features they described, and a few, such as Lewis Evans,[2] organized these speculations into wide-ranging theories.

Among the most important early geological observations in southern travel accounts are those made by Andrew Ellicott in the Mississippi Valley and Florida from 1796 to 1800. Ellicott's geology is not organized into separate chapters in the report of his travels, but is explicit in many widely separated paragraphs or parts of paragraphs and is implicit in many pages. If observations were all Ellicott made, listing them would be useful only to antiquarians. But Ellicott's work is far more important. He was not only a perceptive observer of geomorphic and stratigraphic features, but also sought explanations for their formation. He then organized these to elucidate a part of the geologic history of the Mississippi inner valley and the great limestone plateau that is the peninsula of Florida. He was inductive in his reasoning, proceeding from observations of features, to observations of their formation in time, and then to hypotheses to explain the wider extent of the meander history of the Mississippi and the extent of the organic limestone of the Florida peninsula.[3]

Andrew Ellicott: Engineer and Scientific Observer

Andrew Ellicott (1754–1820) was born in Bucks County, Pennsylvania.[4] During the Revolutionary War he rose to the rank of major, by which title he was addressed throughout his career. His talents in mathematics and astronomy led to a career in precise surveying, and he soon attained a wide reputation. In 1784 he was one of the surveyors who continued the Mason-Dixon Line west from the surveys of 1767. He was appointed to determine the western boundary of Pennsylvania in 1785 and the northern boundary in 1786. He was appointed by the new federal government in 1789 to determine the western boundary of New York. In order to determine a precise point in Canada involving the northern origin of the boundary line, Ellicott took occasion to study Niagara Falls and Niagara Gorge, where he made important geological and hydrological observations. These were published in 1790 and 1799. He accurately measured the height of Niagara Falls and commented on its hydrology. He described the strata over which the river flows and asserted that it was a part of the horizontal strata that extended from the Appalachian Mountains to the Mississippi River and to an unknown distance beyond it. He asserted that the horizontal strata indicated that the "country has never been disturbed by those terrible convulsions which a great part of the globe must have experienced."[5]

One of Ellicott's most important assignments, prior to his work in Florida, was a precise survey of ten miles square in Maryland and Virginia that included the future site of the District of Columbia. The first map of this region was published by Ellicott in 1793. When L'Enfant, the original designer of the capital city, disagreed with the government commission before the work was completed, Ellicott finished laying out the city in 1792 and drew the engraving of the plan of the city. In 1794 to 1796 he surveyed a road through the wilderness of Pennsylvania from Reading to Erie and laid out towns in northwestern Pennsylvania. The government then called upon Ellicott to take charge of the determination of the Florida boundary. His Florida survey will be described in a separate section.

After Ellicott returned from the Florida survey in 1800, he had important appointments in Pennsylvania, Georgia, and elsewhere. In 1813 Major Ellicott went to the Military Academy at West Point as professor of mathematics.[6] He died suddenly in 1820, "with his adored Sally, mother of nine," whom he had married in 1775, by his side.

Ellicott Surveys the Florida Boundary, 1796–1800

Before discussing Ellicott's geological contributions it is appropriate to outline his route to, and along, the Florida boundary and to identify the events that required him to make a precise survey.

The west and east Florida region was Spanish until 1763,

when Spain, having joined France in 1762 in the war against Britain, had to transfer Florida to Britain. The boundary line had been 31°N. latitude from the Mississippi River to the Chattahoochee River, thence south to the mouth of the Flint River, then east (actually south of east) to the headwaters of the St. Marys River in the Okefenokee Swamp, thence down the St. Marys River to the Atlantic. In 1783 Great Britain lost the American colonies and also returned the Florida territory to Spain, who attempted to maintain a more northern boundary and to make Natchez the seat of government. America did not agree to this northern boundary and always insisted on the 1763 boundary of 31°.[7] The contest was settled in 1795, when Spain agreed to the old boundary. It then became necessary for the agreed-upon boundary to be precisely surveyed and properly marked. (Spain ceded Florida to the United States in 1819 and the territory became a state in 1845.)

To cooperate with Spanish officers in surveying the boundary, Andrew Ellicott was appointed commissioner by President Washington, and directed to proceed to Natchez, which was the headquarters of the Spanish government. Ellicott left Philadelphia in 1796 for Pittsburgh, where he secured four boats for his large party. They traveled down the Ohio to the Mississippi and down the Mississippi to Natchez, arriving there in 1797 (Figure 1).

The Spanish governor and his officials, through a series of vexatious actions, delayed the beginning of the boundary survey until May 1798. A less able and determined man would have given up, but through persistent negotiations and a certain amount of private diplomacy, Ellicott finally persuaded the Spanish authorities to join him in beginning the survey.[8] The first step was to determine the intersection of 31° with the Mississippi River. Then he and his three skillful surveyors and their large support party began to mark the line

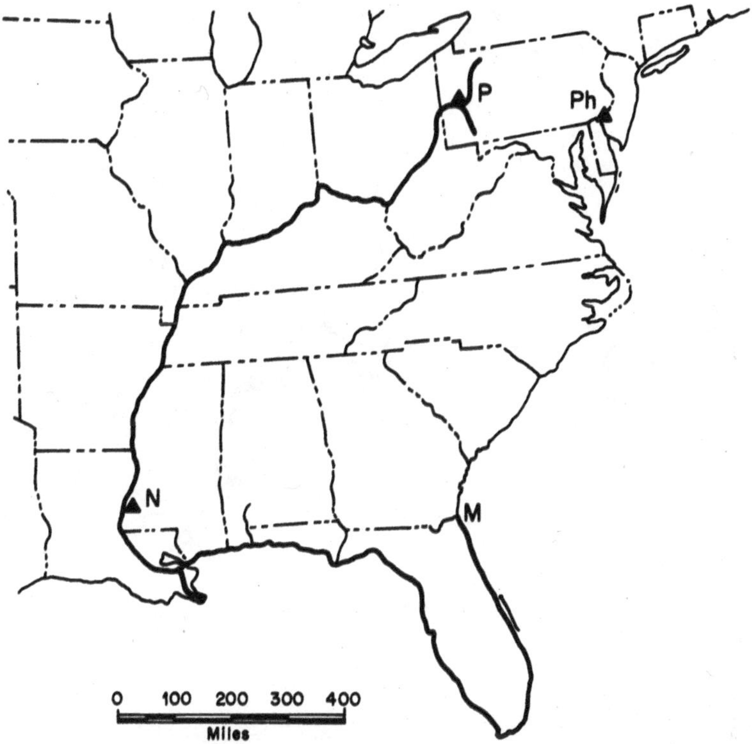

Figure 1.
ELLICOTT'S TRAVELS. Map of southeastern United States to show Andrew Ellicott's travels: Ph *Philadelphia;* P *Pittsburgh;* N *Natchez;* M *St. Marys.*

to the east (Figure 2). At the Pearl River, Ellicott left the party to proceed to New Orleans. Here he purchased an unfinished boat of about thirty-eight tons, which he decked over and rigged as a schooner. He named it the *Sally*, after his

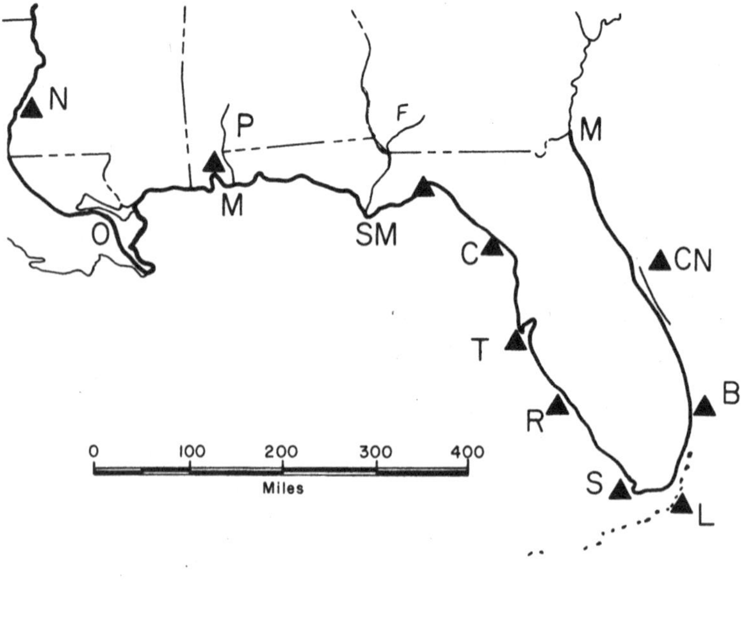

Figure 2.
ELLICOTT'S SURVEY. Localities along Andrew Ellicott's survey in 1798 to 1800 of the Spanish-American Boundary. N *Natchez*, just north of 31°N latitude, which was the boundary to the Flint River; O *New Orleans*, where Ellicott bought a little schooner; M *Mobile;* P *Perdita River;* F *Flint River*, which joins the Chattahoochee River to form the Apalachicola River; SM *St. Marks;* C *Cedar Key;* T *Tampa Bay;* R *Cape Romano;* S *Cape Sable;* L *Key Largo;* B *Key Biscayne;* CN *Cape Canaveral;* M *St. Marys*.

wife. He used this little ship to move supplies from New Orleans to the Mobile, Pearl, Perdida, and Chattahoochee rivers as far as the thirty-first parallel. He was thus able to supervise his surveyors, who met him at these places.

St. Marks, at the mouth of the Chattahoochee (or Apalachicola) River was the principal base on the Gulf. Ellicott himself did not accompany his surveyors from the Chattahoochee to the headwaters of the St. Marys River, but instructed them to proceed without him. He would meet them in St. Marys.

Setting off in the little schooner, Ellicott made his way around Florida to Cumberland Island and the St. Marys River. He had several men in his party, but only two were sailors, men he had found in New Orleans. They were illiterate, however, and could not assist in the navigation of the craft. Apparently Ellicott had little previous experience in taking charge of a sailboat. He was nevertheless an expert mechanic, skilled at precise position-finding, self-confident, and, it is fair to say, a little lucky as well. His report contains maps of the coast only as far as the present northwest Florida, and then of the St. Marys region on the east of Florida. He did not map the shoreline between St. Marks and St. Marys. He had no really precise maps and no charts, except for part of the area of the Keys. The source he used for the names of the bays and islands that he gives in his narrative account is not mentioned, but he may have had Stork's 1769 map of East Florida, as well as others he picked up at St. Marks. There is no evidence that he used Hutchins' 1784 descriptions of coastal features.[9]

In spite of reefs, shoals, and severe storms, which wrecked many other vessels, Ellicott brought his little schooner to St. Marys. His reports of the navigation are almost deliberately casual. He could have used his superb instruments to determine position, but the more simple sextants in his skilled hands were all that were necessary to determine latitude, and apparently only very rough calculations of longitude were needed.

Ellicott did not have "the latitude of St. Mary's on any

publication, but from a small mutilated chart it appeared to be at N 31°." The wind was such that he was carried north of that latitude.

> At four o'clock in the afternoon came to an anchor at the end of St. Simon's Island in a good harbour. The satisfaction which the crew, and myself experienced on this occasion may be more easily conceived than expressed. We were now able to take a night's repose, free from those cares, and anxieties which must ever attend the reflecting mind in our past situation; exposed to the turbulence of the sea in a little vessel, having but two young illiterate sailors on board, along a dangerous coast with which we were all unacquainted, and experiencing three violent gales of wind, which we afterwards found had wrecked as many vessels, much better calculated to resist the fury of the winds, and billows than ours. So great was the dependence on observation, and so little on dead reckoning among the currents near the coast, that the log was never hove once during the passage. (pp. 259–60)

Ellicott cleverly used the Gulf Stream in his passage from the Keys to St. Simons Island. He was familiar with Franklin's essay and maps of the Gulf Stream and took occasion (pp. 261-67) to write at length on the position, width, and speed of the current, and to speculate on its origin.[10]

From St. Simons Island Ellicott returned southward to St. Marys, where he met his party. He and his associates then established an astronomical observatory at Point Peter for position. The reason for his decision not to use the town of St. Marys for this location is of passing interest:

> Finding that I could not obtain quarters in the town for myself and people, free of expense to the public, I removed on the 12th to Point Peter, and encamped in a forest of live oak, where a number of people were engaged in cutting ship timber for the United States: the offal wood served us for firing. This system I pursued from the time I left Pittsburg in the

> year 1796 until my return to Philadelphia in the year 1800; and whatever attention, and shelter the men might require, I occupied no quarters myself at the expense of the public. (p. 270)

After carefully surveying the St. Marys River from its mouth to its origin in the great Okefenokee Swamp, the astronomical part of the survey was finally completed. Ellicott and his party left St. Marys on March 6, 1800, and sailed north to Cumberland Island to complete the maps and reports. Returning briefly to St. Marys, Ellicott chartered a sloop that had come from New York, and set out for Philadelphia May 9. The master of the vessel, having only rudimentary instruments, soon invited Ellicott to take charge of the navigation, which he did with great satisfaction to both. Ellicott took advantage of the Gulf Stream and reached Philadelphia on May 18, 1800, arriving at 8 o'clock in the evening, "when all the fatigues, hardships, and difficulties I had been exposed to during a long absence, were more than compensated by the pleasure I experienced in meeting my family in good health" (p. 300).

In addition to Ellicott's report to the government, he prepared his *Travels of Andrew Ellicott . . .*, which was published in 1803. This book included astronomical tables, maps, the account of his route, and his geologic and other scientific observations and explanations.

Ellicott's Geology in the South

The geological observations with which we are concerned here began at Pittsburgh on October 16, 1796. In the Ohio Valley, they were made at some of the places where Ellicott stopped to determine precisely his position. Exact determinations of position were needed to provide a more accurate map than earlier ones. Hutchins had prepared the first map of the Ohio that was at all accurate, but there were still parts of the Ohio that lacked astronomically determined positions.[11] Ellicott, an accomplished geodetic surveyor, corrected the accepted longitude of the mouth of the Ohio by nearly two degrees (p. 119).

He described the severe air pollution from the smoke at Pittsburgh and the smoke from forest and grass fires from Pittsburgh to Wheeling, which was then a part of Virginia. At times the smoke was so dense that the party "proceeded with difficulty."

Near the mouth of the Scioto River, now Portsmouth, Ohio, Ellicott collected "several fine specimens" of fossils, presumably from the famous Sciotoville bar locality for Mississippian brachiopods.[12]

Across the river in Kentucky he visited salt works and described the salt spring and the manufacture of the salt in some detail (pp. 14, 15). He then described the falls at Louisville and noted the horizontality of the limestones over which the river fell. His description of strata observed in the Ohio Valley is as follows: "The bottom and sides of the river are stony. . . . The strata of stone are horizontally disposed, and principally consist of either freestone, or limestone. This horizontal disposition of the strata of stone, is observable through a very large extent of the United States. I have traced it from Oswego, up Lakes Ontario and Erie, with all waters falling into them, and through all the western part of Pennsylvania, and down the Ohio, whereever hills or mountains are to be seen" (p. 23).

Ellicott continued the map revisions on down the Mississippi from the mouth of the Ohio to Natchez and then to New Orleans and the mouth of the river (*Journal*, plates C, D, and E). He repeatedly commented on the difference between the Ohio Valley, with its hills bordering the river, and the Mississippi, with the floodplain from "thirty-six to forty-five miles wide . . . several feet under water at every inundation."

The formation of meanders of the Mississippi River, their enlargement and eventual extinction, to form what are now known as ox-bow lakes, was analyzed by Ellicott. His own statement and diagram are shown in Figure 3.

Ellicott related the rapid erosion of the concave banks, where "many fatal accidents have happened on this river. . . . When the banks are inundated, they are less dangerous, being in some measure supported by the water, and not so liable to give way" (p. 122). He gave instructions about avoiding perils in the navigation of the river. His scientific acumen and his practical engineering applications are excellently illustrated in this section, as well as elsewhere in his report. He remarked on natural levees: "Some superficial observers, have been led to believe that the river flows along on a hill with a valley on each side" (p. 123).

His wide-ranging scientific interests are well illustrated by his clever determination that the difference in annual mean temperatures between Pittsburgh and just south of Natchez is fourteen degrees. "The permanent degree of heat may be stated at about 14° beyond that of Pennsylvania. The conclusion is drawn from the following facts. In Pennsylvania the mean temperature of the best spring and well water, is about 51° and from the Mississippi, east to the Atlantic, in the parallel of 31° I found it about 65°, the difference is 14°" (p. 135).

It was in West Florida and in the circumnavigation of East Florida that Ellicott made his significant sedimentological and stratigraphic observations and important scientific conclusions. He was not only a keen observer, but also interpreted and explained his data.

At St. Marks, at the head of Apalachee Bay and inland from it, Ellicott "observed a stratum of limestone extending northerly" (p. 241). He observed horizontal limestone all around Florida and finally concluded that all of Florida was a great limestone plateau. The specific locations mentioned are

This river, like all others passing through flat countries, and not checked or confined by hills or mountains, is very crooked as may be seen by the chart. This arises from a very natural cause, and may be explained in the following manner. Suppose in the figure below, the lines *a b*, and *c d*, to be the banks or margins of a portion of a river, and the water moving in the direction *e f*, but meeting with an obstruction

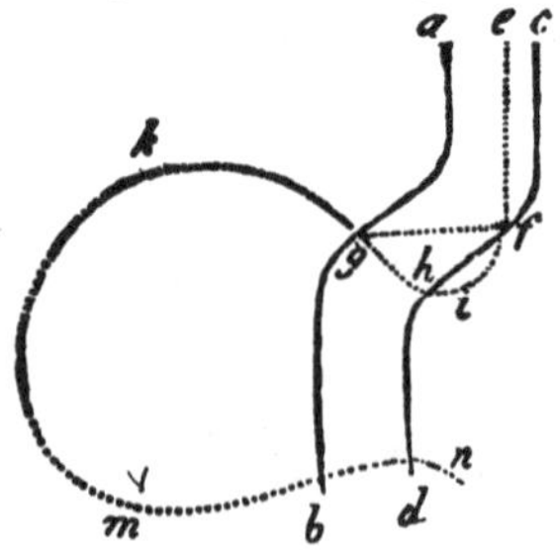

at *f*, it will be reflected in the direction *f g*, and at *g*, as well as at *f*, the bank will be worn away. About *h*, an eddy will be formed, where sand, earth, and rubbish will be deposited, and continually increase the convex parts, while the concave parts will be worn away, and in time a loop will be formed something like the dotted curve line *i*, *h*, *m*, and *n* in the figure, which will increase in magnitude till the river, aided by an inundation, breaks through a shorter way, and the convex part will become an island. If the loop should be very large, and the water cease to have much current along it, the two ends in a short time will be filled up by the great quantity of mud and sand, which are constantly mixed with the water of the Mississippi, and a lake will be formed. These lakes are to be met with in various parts of the swamp, and bear evident marks of having been at some former period, a portion of the main bed of the river.

Figure 3.
MEANDER HYPOTHESIS. Andrew Ellicott's diagram and text to explain formation of meanders and meander cutoffs (from Ellicott, *Journal*, p. 121).

shown on Figure 2. He noted the "horizontal stratum of stone or rock" near Cedar Keys. They passed "Bay Tampe," "Cape Roman," and past Cape Sable to "Sandy Key." Here he noted that "though this island is called Sandy Key, and has certainly the appearance of a body of sand; it is little more than a heap of broken and pulverised shells, which were found to effervesce freely with the vitriolic acid, and little or no quartz was perceptible in the solution" (p. 245).

In a "number of small islands" near "Key Vacos" he extended his repeated observations that the rocks were made up of sand from broken shells. Further, he made the next assumption: that calcareous mud and limestone rock, in which the grain size was so small that no shell fragments could be seen, were really of organic origin. This confirmed the earlier observations of John Bartram, who had traveled in a part of northern Florida in 1765 and 1766 and made some more or less incidental geological observations, as did his son William in 1771–74.[13] Ellicott knew of the Bartrams' reports, and refers to them, but it was to be almost 150 years later that the electron microscope would definitely prove the organic origin of extremely fine-grained limestone.

Ellicott observed the same kind of shell sand and horizontal limestone strata on the east coast of Florida.

> The Florida reef, (as it is called,) appears to consist of a number of coral banks on the outer edge of an extensive stratum of calcarious stone, which extends from the main land, to the edge of the Gulf Stream: the general position of this stratum is nearly horizontal, and is possibly a continuation of that observed at Apalachy. If this should be the case, it may be considered as the base of East Florida, and conform to the general law observed in the disposition of the strata of stone on our western waters.
>
> On this stratum of stone, which serves as a helmet to the southern promontory of East Florida, and defends it from the violence of the Gulf Stream, is situated the whole of that

> cluster of innumerable islands and shoals, which have been so troublesome and dangerous to navigators.
>
> These islands and shoals, may be viewed as protuberances, (standing on the surface of this extensive stratum,) gradually formed during a period of many centuries, by the constant accretion of calcarious matter. Many of those islands and shoals have evidently had their origin from coral banks, which not only like those of oysters, are known to increase, but to surpass them greatly in magnitude: and it is now reduced to a certainty, that a number of the islands in the South Sea are coral rocks covered with a stratum of earth. It is likewise well ascertained by naturalists, that coral is not, as was formerly supposed a vegetable substance, but a vast collection of small animals which build up those rocky edifices from the bottom of the ocean. (pp. 253–54)

Other scientific observations in the *Journal* have more or less relation to geology. With his lifelong military interests Major Ellicott had a keen eye for topography and drainage; he studied these not only to ascertain their geologic origin, but also for their military possibilities (pp. 281–84). He can be regarded as an early ecologist, for he related the topography and drainage to his excellent and interesting account of the animals, especially of the alligators (pp. 277–79). He gave some information on the plants, but referred his readers to Bartram's *Travels through North and South Carolina* for more detailed information. In his consideration of the diseases of the region, he recognized the topographic and climatic controls that account for the differences between tropical diseases and those of the north (pp. 289–97).[14]

Conclusions

Ellicott's geological work in the South in 1796–1800 was undertaken in connection with his detailed and accurate survey of the northern border of Florida, which was until 1819 the southern boundary of the United States. He had already made prescient geological observations, including the gathering of important data at Niagara Falls and in the surrounding region. In his work in the South he extended his concept of widespread horizontal strata throughout the Ohio and Mississippian valleys and the Florida region.

As an engineer he was well aware of the importance of hydrologic and hydraulic action of rivers. He took occasion on his way south along the Ohio and Mississippi rivers to note the mechanism of formation of floodplains, meanders, ox-bow lakes, and natural levees.

One of his important contributions was to determine that Florida's sand was made of broken-up shells, and that still finer particles were also microscopic shell fragments. All the limestone strata of Florida were thus shown to be of organic origin.

Ellicott is an example of a man of the early republic who had a modest formal education, but became skilled not only in mathematics and in the astronomy necessary for a geodetic surveyor, but also along the way became a competent naturalist with geologic ability. He was an important member of the American Philosophical Society and an active member of the Philadelphia "establishment," which included Franklin and Rittenhouse, as well as many others. He contributed articles to the *Transactions of the American Philosophical Society* and furnished material to other scientists.[15]

Ellicott well deserves a place in the history of early American geology. A modern biography giving due attention to his many scientific activities would be a welcome contribution to the history of American science.

Notes

1. Andrew Ellicott, *The Journal of . . . Late Commissioner on Behalf of the United States during Part of the Year 1796, the Years 1797, 1798, 1799, and Part of the Year 1800: For Determining the Boundary between the United States and the Possessions of His Catholic Majesty in America, Containing Occasional Remarks on the Situation, Soil, Rivers, Natural Productions, and Diseases of the Different Countries of the Ohio, Mississippi and the Gulf of Mexico* . . . (Philadelphia: Budd and Bartram for Thomas Dobson, 1803), hereafter cited in the text; J. D. Schoepf, *Beyträge zur mineralogischen Kenntniss des ostlichen Theils von Nordamerika und seiner Gebürge* (Erlangen: J. J. Palm, 1787), facsimile and annotated translation by E. M. Spieker, as *Geology of Eastern North America* (New York: Hafner Publishing Company, 1972); C. F. Volney, *Tableau du climat et du sol des États Unis d'Amérique* . . . (Paris: Courcier, Dentu, 1803), 2 vols., translated by C. B. Brown as *A View of the Soil and Climate of the United States of America* . . . (Philadelphia: J. Conrad and Co., 1804), rpt. 1968, with an introduction by G. W. White (New York: Hafner Publishing Company); William Maclure, "Observations on the Geology of the United States, Explanatory of a Geological Map," *Transactions of the American Philosophical Society* 6 (1809):411–28, issued as a larger book in 1817.

2. Lewis Evans, *An Analysis of the General Map of the Middle British Colonies in America* . . . (Philadelphia: B. Franklin and D. Hall, 1755).

3. G. W. White, "*Birth of American Hydrogeology,*" in *History of Hydrology in the United States,* U.S. Geological Survey Professional Paper, in press.

4. For a broad outline of Ellicott's career see C. V. C. Mathews, *Andrew Ellicott, his life and letters* (New York: Grafton Press, 1908); and the *Dictionary of American Biography,* s.v. "Ellicott, Andrew."

5. Andrew Ellicott, "Description of the Falls of Niagara," *Universal Asylum and Columbian Magazine* 4 (1790):331–32, plate; idem, "Miscellaneous Observations Relative to the Western Parts of Pennsylvania, Particularly Those in the Neighbourhood of Lake Erie," *Transactions of the American Philosophical Society* (1799), pp. 224–29; and George Burgess, *A Journal of a Surveying Trip into Western Pennsylvania, under Andrew Ellicott in the Year 1795* . . ., ed. John Cumming (Mt. Pleasant, Mich.: John Cumming, 1965), p. 226.

6. United States Military Academy, *Annual Report of the Superintendent* (Washington, 1896), pp. 38–43, through the courtesy of K. W. Rapp, Chief, U.S.M.A. Archives.

7. For a contemporary map see Zedidiah Morse, *The American Geography* . . . (Elizabethtown: Shepard Kollock, 1789).

8. For an extended discussion see references in T. D. Clark, *Travels in the Old South: A Bibliography* (Norman: University of Oklahoma Press, 1956), 2:103–6.

9. William Stork, *An Account of East Florida; Third Edition, with a Journal Kept by John Bartram of Philadelphia . . . Botanist to His Majesty for the Floridas; Illustrated with a Map of East Florida* (London, 1769); Thomas Hutchins, *An Historical Narrative and Topographical Description of Louisiana and West Florida, Comprehending the River Mississippi with its Principal Branches and Settlements . . .* (Philadelphia: for the authors, 1784).

10. Benjamin Franklin, "A Letter from Dr. Benjamin Franklin to Mr. Alphonsus le Roy . . . Containing Sundry Maritime Observations," *Transactions of the American Philosophical Society* 2 (1786):294–329; and P. L. Richardson, "Benjamin Franklin and Timothy Folger's First Printed Chart of the Gulf Stream," *Science* 207 (1980):643–45.

11. Thomas Hutchins, *A Topographical Description of Virginia, Pennsylvania, Maryland, and North Carolina, Comprehending the Rivers Ohio, Kenhawa, Sioto, Cherokee, Wabash, Illinois, Mississippi . . .* (London: J. Almon, 1778); see L. A. Brown, *Early Maps of the Ohio Valley . . . from 1673 to 1783* (Pittsburgh: University of Pittsburgh Press, 1959).

12. J. E. Hyde, *The Mississippian Formations of Central and Southern Ohio*, ed. M. R. Marple, Ohio Geological Survey Bulletin no. 51, 1953.

13. John Bartram, "Diary of a Journey through the Carolinas, Georgia, and Florida from July 1, 1765 to April 10, 1766 . . .," annotated by Francis Harper, *Transactions of the American Philosophical Society*, n.s. 33, no. 1 (1942):1–20; William Bartram, *Travels through North and South Carolina, Georgia, East and West Florida . . .* (Philadelphia: James and Johnson, 1791), pp. 80, 314, 315, 378, 399; and idem, "Travels in Georgia and Florida, 1773–74: A Report to Dr. John Fothergill," annotated by Francis Harper, *Transactions of the American Philosophical Society*, n.s. 43, no. 2 (1943): 151,182.

14. For an adverse review of this discussion see Clark, *Travels in the Old South*, p. 105.

15. S. L. Mitchill, "Observations on the Geology of North America . . .," in *Essay on the Theory of the Earth . . .*, ed. G. Cuvier and trans. J. Jameson (New York, Kirk and Mercein, 1818), pp. 319–431. Mitchill not only refers extensively to Ellicott's Florida work on limestones, but also describes fossils from the St. Lawrence Valley near Montreal that Ellicott sent him in 1814.

South Carolina State Geological Surveys of the Nineteenth Century

Anne Millbrooke

As one approach to internal improvement, South Carolina sponsored a series of state geological surveys between the years 1824 and 1860. The support for these surveys resulted in a large part from three related developments: the western expansion of population, the exhaustion of soil, and the competition for western trade. The population of the state and the nation was advancing during this period from the seaboard to the mountains and into the fertile Mississippi River Basin. The then-prevalent methods of cultivation were depleting the soil, causing many farmers on older plantations to emigrate to the West; and trading centers all along the eastern seaboard were vying for trade with the new western settlements. In response, South Carolina sought to improve transportation, as well as agriculture and manufacturing, within its borders. Because of their perceived public utility, geological surveys became part of this response.

At different times Lardner Vanuxem, Edmund Ruffin, Michael Tuomey, and Oscar Lieber directed state geological surveys of South Carolina.[1] Each man conducted his survey with only the occasional assistance of volunteers; none organized a scientific corps to assist in the work. Like many southern scientists, three of the four state geologists studied in schools in the North or in Europe. Vanuxem graduated from the Paris School of Mines. Tuomey trained at the Rensselaer Institute in New York, and Lieber studied in Germany after graduating from South Carolina College. The exception, Ruffin, briefly attended William and Mary College in Virginia. None of the four ever achieved a position of leadership in the American scientific community.

Although the internal improvement movement provided the context in which geology received public support, the creation and continuance of each state survey ultimately rested upon the personalities and politics associated with specific proposals. Consequently, South Carolina repeatedly established geological surveys as limited agencies with practical goals. These surveys exemplify the institutions and uses of science in the antebellum South.[2]

The First Survey

The initiative for the first state geological survey in South Carolina (and the second state survey in the nation) came in 1824 from Lardner Vanuxem, a professor at South Carolina College. Vanuxem filled the newly created chair of geology and mineralogy, the first chair of geology at a state-supported school in the United States.[3] At the time classical studies overshadowed the sciences in schools, and Vanuxem's salary of $1000 was two thirds of that earned by the classics professors. Believing that his salary was "too small to be of any material advantage," he considered resigning.[4] At the urging of the trustees of the college, he instead proposed that the duties associated with a state geological and mineralogical survey be added to his professorial responsibilities and that additional compensation for the survey supplement his professorial salary. This proposal and a recommendation from the college Board of Trustees prompted the state legislature to approve the survey and to appropriate $500 annually for it for two years.

Vanuxem made his survey of South Carolina during school

vacations in 1825 and 1826. He gathered about 500 specimens of minerals and crystalline rocks, which he deposited in the collections of South Carolina College. He also indicated in color on a map the major geological formations of the state. In 1826 he reported on ten kinds of rocks, thirty minerals, their locations, and their value as raw materials. This very brief report covered five districts—Abbeville, Pendelton, Greenville, Spartanburg, and York—in the up country or western part of the state. Although Vanuxem issued only this one report during his tenure as state geologist, he later contributed observations on the state's Tertiary formations in an 1829 article that is significant for its early recognition of the value of fossils in identifying and correlating formations.[5]

After two years, the state discontinued the survey. Vanuxem remained at South Carolina College for an additional year, then resigned to become the superintendent of a gold mine in Mexico. Much of his scientific fame is based on his later work with the New York state survey and its system of geologic nomenclature.

An Agricultural and Geological Survey

By the 1840s several efforts to improve the agriculture of South Carolina coalesced in a movement for another state survey. Reporting on the advisability of establishing a survey, a committee of the House of Representatives concluded "that a Geological and Agricultural Survey of the State, for the examination of our soil, discovery and application of marl lime, and developing all other resources and facilities of improvement, will prove among the most efficient means of giving value to her soil, increasing her products, multiplying her population, and diffusing national and individual prosperity."[6] The committee recommended that a survey be established for a two-year period at an annual cost of $2,000. When brought before the House of Representatives as a resolution, the question called for only an agricultural survey. "Geological" was added by amendment, and the measure passed. In the Senate came a motion to strike the provision for the survey from the general appropriations bill, but the Senate defeated the motion. In 1842 South Carolina thus established an agricultural and geological survey.

Governor James H. Hammond considered only one man for the position of agricultural and geological surveyor: Edmund Ruffin, a noted agriculturist.[7] Years before, while farming an inherited estate in Virginia, Ruffin had become an advocate of agricultural reform, including the protection of crop land from grazing and the use of vegetable manures as fertilizer. Through the years he found such measures inadequate, and undertook experiments to discover the cause of sterile soil. Following a lead in Sir Humphrey Davy's *Agricultural Chemistry*, Ruffin discovered that lime derived from fossil shells neutralizes the acidity of soil and thereby improves fertility. He thereafter promoted the use of marl and other mineral manures, and in 1832 he published his widely read *Essay on Calcareous Manures*. For ten years Ruffin also published a monthly *Farmers' Register*, in which he promoted agricultural improvement, and for a time he served on the Virginia Board of Agriculture.

Before becoming governor, Hammond had attempted privately to employ Ruffin as a consultant on marl, in which Hammond had invested financially. At the time Ruffin had been occupied with other pursuits. In January of 1843, however, Ruffin was temporarily without employment, and he reluctantly accepted the salaried position in South Carolina. Both Ruffin and Hammond were motivated by considerations that went beyond finances; they shared an interest in agricultural improvement and both perceived a religious duty to use their God-given talents for the public benefit. Both men recognized a responsibility to contribute to the South's general welfare.[8]

As the state surveyor, Ruffin traveled about South Carolina, noting the occurrence and promoting the use of marl. Employing a few simple techniques, he examined the plantations of the governor, several legislators, and less prominent planters. He used a long auger to bore through overlying soil, a steel gouge attached to a rod to scoop out samples, and ordinary farm tools to dig holes in overlying soil. He also searched the banks of creeks and streams for evidence of marl. After obtaining samples, he used muriatic acid or vinegar to determine the presence of carbonate of lime, and chemical analyses to define the percentage of lime present. He began the survey in the lower (southern and eastern) districts, where marl and other sources of lime are readily found; later he moved into the upper districts.

Despite his extensive field work, Ruffin believed that his "best service" was done "by talking to & instructing persons" about manures.[9] He met with individual farmers, social clubs, and agricultural societies, and he wrote articles for local newspapers. In addition to promoting the use of mineral manures, he urged farmers to adopt other improved methods of cultivation, like using plows instead of hoes. At the end of the year he prepared the *Report of the Commencement and Progress of the Agricultural Survey of South Carolina* (1843); the state printed 4,000 copies.

Having become ill while conducting the survey, Ruffin returned to Virginia in December of 1843. There he remained, although state officials in South Carolina generally regarded his survey as a success and encouraged him to complete it. In leaving South Carolina, he recommended his friend, the geologist Michael Tuomey, to succeed him as the state surveyor. During the ensuing years Ruffin farmed, defended slavery, and advocated secession. In 1861 he allegedly fired the first shot against Fort Sumter, starting the armed hostilities of the Civil War.

Ruffin's recommendation was accepted, and Michael Tuomey, an Irishman by birth and an experienced teacher and geologist, received the South Carolina appointment.[10] He continued Ruffin's practices of analyzing soils and advising farmers, but he subordinated agriculture to geology. As his first task, he made a geological reconnaissance of the state. Within a year he submitted a brief report in which he identified the main rocks of the state, commented on several metals, and discussed aspects of economic geology. He timed the publication of this report to coincide with the annual legislative session, at which efforts were made to obtain an extension of the survey beyond the original two-year deadline. Agricultural interests provided sufficient support, and the state granted a two-year extension.

Tuomey understood the renewed survey to have five objectives, which he listed:

> 1. The determination and description of the various minerals and rocks of the State.
> 2. Their examination as to extent and relation to each other in their order of superposition, as well as their influence upon the physical features of the State.
> 3. The discovery of metallic veins, and beds of other useful substances, such as lime, rock, marl, &c. that they may contain.
> 4. The relations of the rocks to soils, and their chemical examination, to a certain extent.
> 5. The pointing out of such improvements in mining and metallurgy, as may be thought useful to those engaged in those operations.[11]

Tuomey did not forget the agricultural origins of the state survey, but his main interest was in geology. Furthermore, to reduce the state's almost total reliance upon agriculture, he advocated economic diversification.

Although Tuomey's survey had friends in the state legislature, it did not generate the level of political support that Ruffin's survey had. In addition to the initial two-year extension, however, the legislature granted Tuomey four months, until April 1847, to allow him to complete a final report. Upon completing it, Tuomey moved to Alabama, where he taught at the state university and directed that state's geological survey.

Published in 1848 under the direction of a legislative committee, Tuomey's final *Report on the Geology of South Carolina* contains an introduction to the science of geology, seven chapters on the state's geology, appendixes on related topics, and the first published geological map of South Carolina. It does not contain Tuomey's illustrations of fossils, which the legislative committee considered nonessential. Satisfied with the report, a Senate committee informed the legislature "that the work is not exclusively of scientific character, but with it, has a practical bearing which will make it generally useful to our citizens."[12]

The Lieber Survey

While serving as the state geologist of South Carolina, Tuomey had allowed the teenage son of a college professor to accompany him on field trips. After graduating from college and studying geology in Germany, this youth, Oscar Lieber, campaigned for a new survey of the state. Using the pen name Metallicus, he published a series of articles proposing a survey "of a practical nature—a technical or economic geological survey, whose object shall be to examine and develop all that refers to that portion of the mineral Kingdom within the borders of South Carolina, that can become of practical value."[13] Among the reasons for establishing the proposed survey, Lieber and others in the 1850s mentioned the diversification of industry, discovery of mineral resources, promotion of improved methods of cultivation, and restraint of speculative investments.

When the proposed survey came before the state legislature in 1855, the Senate debated not whether to sponsor a survey, but how to make a survey useful to citizens in every district of the state and at what level to support a survey—to employ one man or a corps of surveyors.[14] Several senators argued that adequate time must be allowed for the completion of a survey, perhaps four years, and that a competent agent must be employed to do the work. These arguments prevailed.

Following the Senate debate, the legislature authorized the appointment of a geological, mineralogical, and agricultural surveyor for a term of four years. Two legislative committees nominated Lieber for the position of surveyor, the legislature duly elected him, and Governor James H. Adams confirmed the appointment—all in December of 1855. South Carolina thus joined the growing number of states that were establishing geological surveys during the relatively prosperous 1850s, when, as a result of the California gold rush, there was a heightened interest in mining.

While Vanuxem, Ruffin, and Tuomey had worked in the natural history tradition of collecting and classifying specimens, Lieber placed his survey in the physical science tradition by using chemistry and natural philosophy to study physical factors. Petrology, the study of rocks, captured most of his attention. To distinguish his work from the paleontologically oriented surveys of other American geologists, including his predecessors, Lieber often declined to call it geological. He preferred the term *geognostic*, meaning that he studied the physical constituents of the earth, and not the manner in which they were formed.[15] On maps, for example, he omitted reference to geological periods and exhibited instead the areas occupied by the different rocks of the state. According to Lieber, geognosy was the chemical and physical study of material substances, of facts, whereas geology was the formulation of theories and laws in an attempt to explain these facts.

During the first year of the survey, 1856, Lieber examined

the Chesterfield, Lancaster, Chester, and York districts. The following year he completed the survey of Union and Spartanburg. In 1858 he investigated the Greenville and Pickens districts; and in 1859, the Anderson and Abbeville districts and part of Edgefield. Although he occasionally worked in the low coastal country during winters, he concentrated on the upper western districts. There mining was a growing industry.

In response to a legislative directive, Lieber tried to discover a limestone suitable for use in the construction of the planned new state house. He concluded that the production of hydraulic cement from a domestic limestone would require "a practice and experience in regard to the individual lime, which would probably for a long time prevent the competition of our article with the imported one, even in our midst."[16] This was the only special assignment added to Lieber's duties as the geological, mineralogical, and agricultural surveyor of South Carolina.

Recognizing the multidisciplinary assignment given to the survey by its enabling legislation, Lieber annually reported on the diverse resources of the state. In each of his four annual reports he described not only the petrology of the districts surveyed, but also the useful minerals and agriculture. Among the minerals discussed were iron, manganese, bismuth, copper, silver, and particularly gold. Limestones, building stones, clays, and coals also received attention. Mining was a major topic, and Lieber provided much-needed instruction in it. He also discussed soils, manures, fencing laws, the mineral consumption of plants, and improved methods of farming. The state published 2,000 copies of each of his reports.

In reviewing the first report, one author remarked that Lieber was "entitled to the highest praise for the creditable manner in which he brought out the result of his observations, albeit the pamphlet report is clothed in the *pea-green-*

literature cover of the news-boys' style. A trifle more expense would have given a durable cover. . . . Now it goes inevitably to the pamphlet heap, with all its fine typography and highly and beautifully finished maps and illustrations."[17] G. Woolworth Colton of Thayer and Colton, a New York firm, engraved the geognostic maps for all the reports. Colton used printed colors instead of hand coloring, printing the maps for the first report from stone, for the second from zinc plates, and for the third and fourth reports from copper plates. Generally, Lieber's contemporaries believed that the theories and conclusions presented in his reports were interesting and should be discussed, but that more information was needed. The reports were of more interest to mining engineers than geologists, who have criticized Lieber's use of "outlandish terms."[18]

The state survey provided Lieber only limited job security. As he explained in 1858, "I am getting to a period of life at which a man has a right to desire some employment that may last him to the end. My present one is not so. When 'the job' is done I shall lie on my oars again."[19] Lieber did not even get to finish "the job" in South Carolina, because the legislature refused to extend the survey. Except for a temporary position with a federal expedition to Labrador, Lieber was unemployed and searching for a job in 1860. When South Carolina seceded from the Union late that year, he enlisted in the Confederate Army. Although continually petitioning for a promotion and transfer to the topographical division of the engineering department, he served as a private in an infantry unit. He died from a wound received at the Battle of Williamsburg.

The Civil War effectively ended South Carolina's efforts to support geology during the nineteenth century. In 1865 Union troops occupied the South Carolina College building that housed the state's geological and mineralogical collections, including specimens from the several state surveys.[20]

The troops threw the collections onto the lawn, where Joseph LeConte, professor of chemistry and geology, and James Woodrow, president of the college, salvaged what they could. Not until February of 1901 did the state approve legislation establishing another geological survey.

Notes

1. The surveys are discussed in J. A. Holmes, "Mineralogical, Geological and Agricultural Surveys of South Carolina," *Journal of the Elisha Mitchell Scientific Society* 7 (1890):89–117; George P. Merrill, *The First One Hundred Years of American Geology* (1924; New York: Hafner Publishing Company, 1964), pp. 101–2, 239–40, 264–66, 323–25; and Max Meisel, *A Bibliography of American Natural History* (1924–29; New York: Hafner Publishing Company, 1967), 2:437–41.

2. Regarding science in general in the South, see Thomas Cary Johnson, Jr., *Scientific Interests in the Old South* (New York: D. Appleton-Century Company, 1936); and Clement Eaton, *The Mind of the Old South*, rev. ed. (1964; Baton Rouge: Louisiana State University Press, 1967), pp. 224–44.

3. Charles S. Sydnor, "State Geological Surveys in the Old South," *American Studies in Honor of William Kenneth Boyd*, ed. David Kelly Jackson (Durham, N.C.: Duke University Press, 1940), p. 90.

4. November 24, 1824, U.S.C. Board of Trustees Minutes, vol. 2, 1813–1837, in the South Caroliniana Library, University of South Carolina, Columbia. An account of Vanuxem's appointment appears in M[aximilian] LaBorde, *History of South Carolina College, from Its Incorporation December 19, 1801, to Nov. 25, 1857, Including Sketches of Its Presidents and Professors* (Columbia: Peter B. Glass, 1859), pp. 143–48. Regarding Vanuxem, see also the *Dictionary of American Biography* and the *Dictionary of Scientific Biography*, s.v. "Vanuxem, Lardner."

5. Lardner Vanuxem and S. G. Morton, "Geological Observations on the Secondary, Tertiary, and Alluvial Formations of the Atlantic Coast of the United States of America," *Journal of the Philadelphia Academy of Sciences* 6, pt. 1 (1829):59–71; see also Lardner Vanuxem, "Report on a Mineralogical and Geological Examination of the State of South Carolina," *Statistics of South Carolina, Including a View of Its Natural, Civil, and Military History, General and Particular*, comp. Robert Mills (Charleston: Hurlbut and Lloyd, 1826), pp. 25–30.

6. House of Representatives, Committee on Agriculture, "Report," *Re-*

ports and Resolutions of the General Assembly of South-Carolina. Passed at Its Regular Session of 1842 (Columbia: A. H. Pemberton, 1843), p. 92.

7. See the Hammond to Ruffin letters for 1842–43 in the Edmund Ruffin Papers, Southern Historical Collection, University of North Carolina Library, Chapel Hill. Sketches of Ruffin appear in the *Dictionary of American Biography;* "Edwin *[sic]* Ruffin, of Virginia, Agriculturist, Embracing a View of Agricultural Progress in Virginia for the Last Thirty Years," *De Bow's Review,* o.s. 11 (1851), pp. 431–36; and Earl G. Swem, "An Analysis of Ruffin's Farmers' Register, with a Bibliography of Edmund Ruffin," *Bulletin of the Virginia State Library* 11 (1918): 40–144. See also Edmund Ruffin, "Incidents of My Life, 1821–1851," MSS 1 R 8385 a 2, Virginia Historical Society, Richmond.

8. Drew Gilpin Faust, *A Sacred Circle: The Dilemma of the Intellectual in the Old South, 1840–1860* (Baltimore: Johns Hopkins University Press, 1977).

9. Entry of March 8, 1843, "Private Diary of Ed. Ruffin, Agricultural Surveyor, in S. C.[a] 1843," MSS 1 R 8385 a 1, Virginia Historical Society, Richmond.

10. Eugene A. Smith, "Sketch of the Life of Michael Tuomey," *American Geologist* 20 (1897):204–12; William B. Rogers, Tuomey obituary, *Proceedings of the Boston Society of Natural History* 6 (1857):184–86; and "Prof. Tuomey," *American Journal of Science* 73 (1857):448.

11. M. Tuomey, *Report on the Geology of South Carolina* (Columbia: A. S. Johnston, 1848), p. ii; see also idem, *Report on the Geological and Agricultural Survey of the State of South Carolina* (Columbia: A. S. Johnston, 1844).

12. Senate, Committee on Agriculture and Internal Improvements, "Report," *Reports and Resolutions of the General Assembly of the State of South Carolina, Passed at the Annual Session of 1848* (Columbia: I. C. Morgan, 1848), p. 204.

13. Metallicus [Oscar M. Lieber], "The Technico-Geological Survey of South Carolina," *Charleston Mercury,* November 16, 1855; see also the articles in the October 24, 1855, and October 25, 1855, issues of the *Charleston Mercury*. A sketch of Lieber appears in the *National Cyclopaedia of American Biography,* s.v. "Lieber, Oscar M."

14. "Geological Survey," *South Carolina Legislative Times,* 33rd Legislature, 2nd Session (December 11, 1855), pp. 171–73.

15. Oscar M. Lieber, *Report on the Survey of South Carolina: Being the First Annual Report to the General Assembly of South Carolina, Embracing the Progress of the Survey during the Year 1857; with Plates and Maps,* 2nd ed. (1857; Columbia: R. W. Gibbes, 1858), p. 5. The natural history

and physical science traditions are discussed in Nathan Reingold, *Science in Nineteenth-Century America: a Documentary History*, American Century Series (New York: Hill and Wang, 1964), pp. 162–63, 59–62.

16. Oscar M. Lieber, *Report on the Survey of South Carolina: Being the Second Annual Report to the General Assembly of South Carolina, Embracing the Progress of the Survey during the Year 1857; with Plates and Maps* (Columbia: R. W. Gibbes, 1858), p. 121.

17. [Thomas McElrath], review of Oscar Lieber's first annual report, *Mining and Statistic Magazine* 10 (1858):175.

18. Merrill, *First One Hundred Years*, p. 324; see also the review of Lieber's third annual report, *American Journal of Science*, 2nd ser. 28 (1859):148.

19. Oscar M. Lieber to parents, December 11, 1858, Francis Lieber Papers, in the South Caroliniana Library, University of South Carolina, Columbia.

20. David Strobel Martin, "The Mineralogical and Geological Cabinets of South Carolina College," *South Carolina College Bulletin* 4 (1906):11–22.

Early American Geological Surveys and Gerard Troost's Field Assistants, 1831–1836

James X. Corgan

In the 1820s and 1830s a new group of scientists appeared on the American scene. These were the field naturalists, who staffed state-supported geological surveys. The first state survey was established in North Carolina in 1823,[1] and the next was authorized by the South Carolina legislature a year later. In 1830 Massachusetts followed suit, and Tennessee joined the movement in 1831.[2] Each pioneer survey was a small-scale operation, a second job for one hardworking college professor.

Despite their part-time, low-budget status, the early geological surveys received a great deal of political attention, because they offered very attractive prospects for both short-term and long-term economic gains. Field workers who staffed the surveys were believed to be competent to analyze soils, to recognize local sources of mineral fertilizers, and to offer sound advice on soil-fertilizer relationships. Further, field workers could guide water-power development, identify minerals, classify rocks, choose rights-of-way for road and canal construction, appraise ores, locate new mineral deposits, and give wise counsel to the mineral industries. With suitable prodding and careful legislative control, field workers also wrote attractive little books that considerate legislators could distribute free of charge to their constituents. The state got a great deal of political and economic benefit from the early surveys and the monetary investment was trifling. Some part-time geologists received as little as $250 or $500 per year.

Modern historians recognize these early one-man, state-financed ventures as part of a continuum that leads through time to modern, scientific endeavors like soil surveys, bureaus of mines, and geological surveys. The one-man surveys are in turn the direct descendants of the mining departments and similar agencies that Giovanni Arduino and other European scholars worked for, generally on a part-time basis, in the eighteenth century.[3] By the middle 1830s American state geological surveys began to change character. By 1833 Maryland had a survey with two employees. In 1836 Pennsylvania authorized a survey with four professional positions. Virginia, New York, and Ohio followed Pennsylvania's lead, and soon surveys with large staffs became the norm. By about 1837 geology had emerged as a recognized profession and the geologist was one of the various sorts of professional people that most state governments employed.

At the same time that the science of geology was developing a professional identity, important semantic changes were also taking place. In the early 1830s and in prior decades the science that we now called geology lacked a universally recognized and consistently used name.[4] Some people used the term *geology*, but to some it was simply mineralogy, and others preferred the term geognosy. Further, this science without a universal name was a poorly defined discipline. In addition to all of modern geology and much of modern soil science, it included medical topography, parts of physical geography, and elements of both civil and mining engineering. Although the geology of the 1820s and early 1830s was a loosely structured discipline, it was a science that served the public in meaningful ways, and it soon emerged as a recognized professional field. By the late 1830s the term *geology* was almost universally used for this polymorphous science, and another important word was entering the vocabulary of English-speaking people. The word *scientist* was coined in 1834.[5] While this new word did not gain immediate acceptance, by the 1840s most field workers who participated in state geological surveys thought of themselves as scientists, as geologists, and as practitioners of a profession.

In the early decades of professionalization in American geology, the way in which maps were made and the things that a geologist did in the field varied significantly from one geographic area to the next. Some concept of the magnitude of this variation can be obtained by comparing articles within the present volume. The emerging science of geology was in a state of flux and the few field workers who were trained before the beginning of this period of experimentation were in a unique position to influence the development of American field traditions.[6] The present study examines the accomplishments and tries to appraise the influence of one group of these preprofessional field workers, the people who worked on the Tennessee geological survey in the years before 1837.

Gerard Troost (1776–1850)

The director of the first Tennessee state geological survey, Gerard Troost, was born in central Holland in 1776 and emigrated to the United States in 1810. When he came to America he was the only man in the country who had ever earned a living through the professional application of geological knowledge. From July 1807 through March 1810 Troost had been employed by the influential Louis Napoleon as keeper of a rather large collection of minerals. For most of this time, Louis Napoleon was king of the Netherlands, and Troost was involved in building a royal cabinet. As Louis

Figure 1.
GERARD TROOST (1776–1850). From a deteriorating photograph in the private collection of Miss Margaret Lindsley Warden, of Nashville, used with her permission. The photograph was probably taken about 1848.

Napoleon's representative, Troost lived in Paris and traveled widely.

During his Parisian years, Troost was also a scholar in residence at the museum of natural history. Within the museum he was the protégé of Haüy, the founder of crystallography, and he was also part of an intellectual circle that included many truly eminent scientists, men like Cuvier, Lamarck, and von Humboldt.[7] Troost, in essence, served an advanced academic residency in one of the world's foremost scientific research facilities, and he arrived at a most opportune time. From 1807 through 1811 the museum was unusually well funded and relatively tranquil.[8]

Much that has been written about Troost, especially about his European background, reflects studies by L. C. Glenn and by H. C. Rooker, studies that were made in the early twentieth century without the extensive use of manuscript sources.[9] It seems wise to question their undocumented statements and to disregard interpretations of Troost's motivation. From sources that reflect research in European repositories, it seems clear that Troost had practiced medicine and that he had strong academic credentials in pharmacy.[10]

Troost's arrival in North America in 1810 created at least a mild stir within the scientific community. He came with letters of introduction from Haüy, and perhaps from other European intellectuals. He also brought with him identifications of North American minerals that had been sent to Paris for expert study.[11] From the time of his arrival in the United States until 1831, when he became state geologist of Tennessee, Gerard Troost was continuously involved in the evolution of American science and was frequently engaged in geological field work. For one thing, he developed one of the finest private mineral collections in North America, based at least in part on personal collecting. The manuscript catalog of his collection documents numerous brief field excursions to collecting areas in the eastern United States.[12] Clearly, most trips were made before 1825.

The most important of Troost's early excursions was a twenty-seven-day trip through New Jersey, made in 1816. The organizer of this venture was William Maclure, justly called the Father of American Geology. Like Troost, he had acquired much of his geological knowledge in Paris. Maclure and Troost traveled with Charles Alexander Lesueur, who is treated at length in a later portion of this narrative.[13] The trio discussed American geology in French, since Lesueur was new to American culture and spoke no English.

In the mid-1820s Troost was also involved in two large-scale field projects: mapping the geology of the Philadelphia area and examining the mineral lands of the American frontier. In each project he made extensive use of a field assistant. For work in Philadelphia, Gerard Troost's assistant was his younger brother Benoist Troost (1786–1859).[14] The younger Troost was a Paris-trained physician-naturalist who preferred science to medicine. He moved from Java to Philadelphia in 1810, arriving at about the time his older brother migrated from Europe.

In their pioneer mapping of the Philadelphia area, the Troost brothers had a shortage of local models to guide their development of a geological map. A recent and magnificent bibliography of early American geology suggests that in 1824 less than a dozen detailed geologic maps had been published for North American localities.[15] Most were inelegant. The Troost brothers were two Europeans trying to do something novel, and for the novel and the innovative, their frame of reference was continental, basically French.

Long before the Philadelphia map was published, Gerard Troost moved to the American frontier. He settled in New Harmony, Indiana, probably in 1825.[16] In New Harmony, Troost was part of a utopian socialist community financed by two widely known men of wealth, Robert Owen and William Maclure. New Harmony was to be the site of a great social, cultural, and economic experiment. Most of the intellectuals

BENOIST TROOST (1786–1859). From an oil painting made in 1859 by George Caleb Bingham. The original is in the Nelson Gallery–Atkins Museum, Kansas City, Missouri, which granted permission for its reproduction. It was a gift of the Kansas City Board of Education.

involved in the New Harmony venture had previously lived in the Philadelphia area. Many had their roots in French culture, and most of them were economically dependent upon the patronage of William Maclure. While Troost apparently was financially independent, he was part of Maclure's contingent.

Far out on the geographic frontier, Maclure's urban intellectuals tried to cross a major cultural frontier by designing the perfect school system. In the first stage of implementing the perfect system, the highest school was the Adult School, where Gerard Troost was to be the teacher of mathematics and science. Although the prestige of this position must have been attractive to him, Troost came to the frontier for personal reasons. A widower with two children, he had just remarried. Further, one of his major business interests, a mining and pharmaceutical company, had recently failed. A new start beckoned, and for anyone in the American intellectual community the Maclure-Owen experiment was regarded highly as a great opportunity to achieve prominence as a scholar.

On February 26, 1826, Troost and Charles Alexander Lesueur set out from New Harmony to examine the known mineral deposits of the frontier. Lesueur was an old friend of Troost's and a naturalist of exceptional experience. Summaries of his French-language notes on his travels in North America have been published several times, most recently in English translation.[17] Troost and Lesueur spent about two months in the field. While they did not produce a map, they did publish several mineralogical articles, articles that are not important in the present context. What is important is that Lesueur never mastered English. Troost was, once again, doing field work in a French cultural setting.

When he returned to New Harmony, Troost faced serious problems. The Maclure-Owen adventure was failing and the intellectual community was beginning to implode. Within the French-speaking group, factions were developing. Gradually,

Figure 3.
CHARLES ALEXANDER LESUEUR (1778–1846). From a sketch by Karl Bodmer, probably made in 1832. This illustration was copied from a print in the collection of Historic New Harmony, Inc.

Troost was drawn into the fray. His situation worsened when his religious wife did not get on well with women who followed other ways of living. By 1827, hostility and name-calling had reached the point where Troost's honesty was questioned. Then, on top of everything else, his fifteen-year-old daughter eloped.[18] It was time to move.

During their travels in 1826 Troost and Lesueur had met a young friend from Philadelphia, George Bowen. Bowen, a promising mineralogist, was teaching at the newly organized University of Nashville. Perhaps because of Bowen, perhaps for other reasons, Troost left New Harmony in the summer of 1827 and settled in Nashville. He soon became proprietor of a museum and then started teaching at the University of Nashville.[19] In the winter of 1830–31 he was quite certain that he would be appointed state geologist. His friend Lesueur was, once again, available as an unpaid field assistant. Lesueur arrived in Nashville in March of 1831 and the two thoroughly experienced field workers began a new adventure, mapping the geology of Tennessee.[20] Lesueur, who is famed for his artistry, made drawings for an anticipated state report, which was never published,[21] and Troost began to prepare a map, which was published in 1835.

In the spring of 1831 Lesueur returned to his own career. Benoist Troost was also preoccupied with his own affairs. For the first time in his years in the United States, Gerard Troost was left without the services of a properly trained, French-speaking field assistant. Troost's need for a traveling companion and assistant grew more urgent in December of 1831, when the state legislature finally approved his appointment as state geologist. It was a great opportunity and a great challenge to a field naturalist, but the gregarious Dr. Troost liked to work with an assistant. As a professor at the University of Nashville he was in contact with many students who found science interesting. From these students and from other sources he soon developed a corps of eager young field apprentices, all of whom worked without pay.

Today it is difficult, almost impossible, to list the names and identify the contributions of each person who labored in the field with Troost. State documents of the era are generally bare-bones accounts, lacking such frivolities as acknowledgments, and Troost's style of work made acknowledgments especially difficult. His studies were generally regional surveys in which he stressed synthesis, rather than local detail. He built up knowledge gradually. State reports that were published in the 1840s often reflected field work that was essentially completed in the 1830s but had been reevaluated on several later occasions. Different field assistants worked with Troost in the same area at different times and the same assistants worked in different areas. It seems impossible to work out an exact chronology of Troost's geological excursions. Still, there are a few sources of insight, and a few sources of more exact information on Troost's relationships with younger, field-oriented scientists.

Three of Troost's assistants are here studied in some detail. They were among his earliest protégés who spoke English as a mother tongue. Each went to the field with Troost, and each later built a scientific career that involved extensive professional-grade geological field work.

Abram Litton (1814–1901)

The name of the Irish-born Abram Litton is virtually unknown in the annals of American science, yet he was both a pioneer in chemical education and a central figure in the first wave of geological field work in several frontier states. Litton was eighty-seven when he died in St. Louis in 1901. For most of the last five decades of his life he held two full-time, or near full-time, academic appointments: professorships of chemistry in geographically proximate universities. Understandably he was a man of modest scholarly attainments, but as a youth he was a person of extraordinary promise.[22]

Figure 4.
ABRAM LITTON (1814–1901). From a portrait of unknown origin that accompanied each of the published biographies of Litton that are discussed in note 21. Efforts to locate the original portrait were unsuccessful.

The relationship with Troost began in 1829, when young Litton entered the University of Nashville with sophomore standing. He was fourteen. After graduating in 1831, Litton spent an additional year in residence at the university and then moved to West Tennessee as an academy-level teacher. He also began to study medicine with a preceptor, but did not complete his apprenticeship. In 1835 he returned to the University of Nashville as professor of mathematics and natural philosophy.[23] The University had a faculty of four, including the president. As Troost was the senior professor, employing a former student to join the faculty in his area of expertise probably required Troost's approval.

In 1838 Litton resigned his position and left for study in Europe. His focus of interest was chemistry, a subject Troost taught. After a brief visit to Great Britain, Litton settled in Paris, probably in July of 1838.[24] The institution where he first studied European advances in chemistry is at present unknown. Litton was also interested in other aspects of European science, especially in the scientific side of agriculture. He made an effort to acquaint himself with French governmental agencies and other institutions that dealt with agricultural education. Eventually he was author of a brief note in an American journal describing French efforts in agricultural education.[25]

In 1839 Litton grew dissatisfied with laboratory facilities in Paris and decided to search for a more suitable place to study chemistry. He walked through southeastern France and Switzerland, visiting a famed agricultural school at Hofwyl. Then he walked on to Heidelberg, visited briefly, and continued through Bonn to Giessen, where the prominent agricultural chemist Justus Liebig had his laboratory. Litton arrived in Giessen in 1839, probably in June, although the month is far from certain. He spent about six months with Liebig and then moved to Berlin, where he spent about half a year with Rosa. He then took a vacation in Switzerland before returning to

Germany to study for almost a year in the laboratory of Wöhler.

In 1842 he returned to Nashville, married, and moved to St. Louis, where he began to teach chemistry in the St. Louis Medical College. Soon after he became a chemistry professor, Litton published the introductory part of his lecture notes.[26] They are, sadly, impersonal and provide no data on his travels in Europe.

Litton's years as a vagabond student are seemingly unrecorded in the rather extensive literature on Americans who studied chemistry in Europe during the nineteenth century. To cite an example, Litton is not mentioned in a recent, excellent, and thoroughly documented list of Liebig's non-German students.[27] Yet all of Litton's biographers agree that he was Liebig's student. Further, one of his letters to Liebig survives in English translation.[28] It is easy to understand how a Liebig-Litton relationship might have been overlooked. Litton was, apparently, a taciturn man with extraordinary linguistic skills. He blended into the European population and he came to Giessen without prearrangement. He was an ultrabright, offbeat traveler who may never have enrolled for anything.

After settling in St. Louis in 1842, Litton spent his school vacations in Tennessee, where he joined Troost in the first detailed geological mapping of the Central Basin. Litton's participation in this major mapping project can be documented by a few very meager notes in the intellectual journals of middle Tennessee, journals that did not exist before the 1840s.[29] Despite a total lack of documentary proof, it seems certain that Troost and Litton worked together in the field from the earliest days of their relationship, from the days when Litton was a boy in his teens. There are a few vague indications of a close relationship. Sometime in the 1830s Litton became fluent in French and German, languages that Troost spoke. Troost and Litton were once teacher and student but eventually became equals, two science teachers at a tiny university. Finally, when Litton went to study in Europe, he began in the city where Troost had studied. It is reasonable to conclude that Troost influenced Litton and that they were associated, both on campus and in the field, for a long, long time.

In the late 1840s opportunities for frontier geological mapping increased and Litton joined in a pioneer survey, the federally supported survey of the Chippewa Land District.[30] It should be emphasized that Litton was not the director of this survey, and indeed, leadership would have been totally out of character. The survey was divided into five mapping teams, with a total staff of twenty-six. Litton headed one team.

Chippewa land holdings were immense, stretching from the Mississippi to Lake Superior and overlapping many modern states. Mapping took five months, but it is not certain how much of the Chippewa domain was explored by Litton and his crew. He probably examined a large area in Wisconsin and smaller tracts in Minnesota and Iowa. Litton became very deeply involved in the mapping project, moving his family to New Harmony, Indiana, where the survey was headquartered.[31] He may have helped with laboratory analyses, but this involvement is not clearly recorded. Late in 1847 Litton and his family returned to Missouri.

In Missouri there was talk of a state geological survey. Eventually such an enterprise was funded, and on April 16, 1853, Litton became the first assistant state geologist and the first state chemist. He was reappointed for an additional two years beginning on February 16, 1855.[32] As he had done in the Chippewa Survey, Litton held a responsible position, but he remained in the background. The state geologist, George C. Swallow, may have been influenced by the well-traveled Litton, but this cannot be documented.

To summarize, from the late 1820s through the early 1840s

Litton developed an extraordinarily rich background in the sciences. His laboratory skills were developed in the best European institutions. His field skills were initially acquired from Troost in Tennessee and then embellished by extensive travel. Litton accepted appointments as a geologist and as a chemist on government-financed geological surveys. What he did as a geologist and what he did as a chemist remain somewhat mysterious, probably because he was a retiring man who wrote very little and never sought a role of leadership in the sciences. Still, he was very knowledgeable and may have been a person from whom less experienced colleagues could learn.

David Dale Owen (1807–60)

The scientist in charge of the Chippewa survey, where Abram Litton was employed, was David Dale Owen. Owen and Litton were friends of long standing, two protégés of Gerard Troost who differed totally in personality. While Litton was content to remain obscure, Owen sought prominence. He was an innovative, resourceful, creative scientist who communicated effectively with political leaders, scholars, and the common man. Much has been written about the socially, economically, and intellectually prominent Owen.[33]

An Owen-Troost relationship began to develop long before the two men actually met. Owen's father, Robert Owen, was a cofounder of the New Harmony community, where Troost had taught, and they had many acquaintances in common. Troost, however, left Indiana before David Dale Owen made New Harmony his home.

Owen became a medical student at the Medical College of Ohio, attending in 1835–36 and 1836–37.[34] Although he was ostensibly training for a medical career, he was actually preparing for a career in the sciences. He was personally

Figure 5.
DAVID DALE OWEN (1807–60). From a sketch made in the late 1840s and first published in Owen's federal report on the Chippewa survey. This sketch has been reproduced in most published biographies of Owen.

acquainted with many scientists and he was also well read. For many years chemistry had interested him strongly, but in the mid-1830s he veered toward geology. Perhaps Maclure's philosophy of geology influenced his decision; perhaps he was swayed by a close friend, Henry Darwin Rogers, who had entered geology as a profession. Whatever the cause, before his first year of medical school ended, Owen had decided to become a geologist.

In 1836, during the break between academic years in medical school, Owen traveled to Nashville to learn geological field skills from Gerard Troost.[35] Then, in 1837, when Owen received his medical degree, he plunged into a truly unique career. From a base of operations in New Harmony, Owen conducted what was, in essence, a nationwide consulting practice in field geology. His clients were occasionally private companies, but his typical business relationships were with governments. He contracted to perform geological field surveys of states or of federally owned territories that were the size of states or larger.

In all the history of geology, David Dale Owen is the only person who ever specialized in this type of service. One might argue that he became the foremost field geologist of all time, *the* frontier mapper. Modern states that lie at least partially within areas that David Dale Owen surveyed include Minnesota, Wisconsin, Michigan, Iowa, Arkansas, Indiana, Kentucky, and Illinois. In addition, Owen or members of his field crews touched on the geology of contiguous areas, chiefly in West Virginia, Ohio, South Dakota, North Dakota, Nebraska, and the plains of western Canada.

It would be wrong to imply that Owen's geological career was an extension of Troost's work. Owen produced much more than Troost, and he was influenced by many other people. He was his father's son, strongly influenced in his youth by Robert Owen's social thought. He may have been very strongly influenced by the geological philosophy of his father's former partner, William Maclure. The fact remains, however, that David Dale Owen was a strong, independent, innovative person who developed his own distinctive approach to managing a geological survey. Owen was primarily a field worker. For the field-oriented aspects of geology his frame of reference, his point of departure for innovation, his nascent impressions of procedure were grounded in his experience in 1836, when he served an outdoor apprenticeship under Gerard Troost.

Through Owen, Troost's influence on the field worker's craft extended into the frontier territories of the north and the west. Owen, in turn, trained a legion of field workers who got their initial exposure to outdoor employment on the surveys he directed. He also offered intense training to many who settled in New Harmony to work with him on a continuing basis. This aspect of Owen's impact on American geology has recently been discussed by Markes E. Johnson.[36] Through the men Owen trained, a complex heritage was passed forward. Part of that heritage may have originated with Owen, some of the philosophy probably came from William Maclure, and some of the data-gathering skills may have been Troost's. Through Owen, some small vestige of Troost's field procedures surely permeated most North American geological surveys of the later nineteenth century.

Richard Owen Currey (1816–65)

David Dale Owen, Abram Litton, and Charles Alexander Lesueur knew each other and were at least temporary members of the Nashville scientific community in the early to middle 1830s. One other member of that community must be discussed in appraising Troost's influence on geological field work in North America. Richard Owen Currey was a native Nashvillian who became Gerard Troost's most loyal disciple.

Unlike Owen, Lesueur, and Troost, Richard Owen Currey

Figure 6.
RICHARD OWEN CURREY (1816–65). From a glass negative in the First Presbyterian Church Collection of the Tennessee State Archives, Nashville, which granted permission to reproduce it. The original of the portrait, which was probably made about 1849, could not be located. This is the first published illustration of Dr. Currey.

is an unknown figure in the history of American science, even more obscure than Abram Litton, who at least became the subject of two fairly conventional biographical notes. Thus far, the only published commentaries on Currey's career are very brief and quite narrowly focused: one appraises his role in the history of medicine; another assesses his influence on the history of geology. The only other biographical fragment is a death notice, just thirteen lines long, that appeared eleven years late.[37] In one biographical discussion Currey's name is misspelled, a rough indication of the quality and quantity of intellectual labor that has gone into an evaluation of his career.

Despite a failure to achieve posthumous fame, during his lifetime Currey was a well-known leader in the southern scientific community. He was founding editor of three scientific journals, an innovative college professor, an organizer of many scientific groups, and a prolific author.[38] Currey's present obscurity can be traced to the environment in which he died; he suffered a routine death in a prisoner-of-war camp in 1865.[39] If 1865 had been a normal year, obituaries in learned journals would have catalogued Currey's many accomplishments, and perhaps some loyal student would have written a book-length biography. But 1865 was not a normal year, and in the years after 1865 southern intellectual affairs were so adversely affected by the Civil War that most ties with the prewar scientific culture were broken. By the late nineteenth and early twentieth centuries, when new scientific societies and new learned journals evolved, Currey's old journals and the intellectual organizations of the antebellum era were largely forgotten. Thus, Currey has remained unknown.

In order to document Currey's field-based contributions to geology, it is necessary to examine his career in some detail. Unlike Troost's other assistants, Currey was never employed by a state or federal agency, but he did produce many privately published reports that were equivalent in scope and in quality to the best state documents of the antebellum era. Currey's magnum opus was *A Sketch of the Geology of Tennessee*, published in book form in 1857 and presaged by a host of shorter studies that can be traced all the way back to a medical dissertation he submitted in 1840.[40] In the introduction to his *Sketch*, Currey's biographical comments suggest he first became involved in field work with Troost about 1836, the year he graduated from the University of Nashville, where Troost had been his professor. They apparently remained on close terms and traveled together occasionally until Troost died in 1850.

As an independent geological mapper, working for the fun of it, Currey produced some masterpieces, most notably a small tome on the copper region of Virginia.[41] This publication was accompanied by a detailed geological map, Currey's best cartographic effort. Both the book and the map were republished in 1880.[42] They probably continued to be used well into the twentieth century. Currey also wrote many routine but journeyman-grade reports. To cite one example, he published an overly brief but perceptive comment on the structure and mineral deposits of northern Alabama.[43] In this study he did not produce a map and he did not carry his work beyond the reconnaissance level.

Currey worked in three states: Virginia, Alabama, and Tennessee. Most if not all of Currey's well-known field studies were blended into publications of late nineteenth and early twentieth century geological surveys.[44] Thus data have been conserved and the work of Troost's disciple Currey continues to have some mild influence on contemporary geology.

Nebulous Relationships

In the early 1830s few Americans had done extensive

geological field work, and thus few could offer experience-based instruction in the field worker's craft, which is complex. Twentieth-century scientists know the difficulties of keeping good field notes, setting priorities in the search for data, and using field instruments and maps. With more primitive instruments and cruder maps, data gathering must have been more difficult in the 1830s. Yet the field workers of the era produced maps and diagrams that chart both the structure of the earth and the sequence of strata in a meaningful way. Today, learning to do field work requires at a minimum six to eight weeks in a formal field camp, plus appropriate background courses. In the 1830s, tagging along with an expert for two weeks or a month apparently provided an equivalent degree of training. Litton and Currey learned from Troost as part of an extended professor-student relationship. Owen learned because he planned to specialize in field studies. There were at least four other people who learned, or may have learned, field skills from Troost during the early 1830s.

Elsewhere in this volume Martha Coleman Bray discusses southern influences on the scientific career of Joseph Nicollet, who spent six weeks with Troost in 1832. Nicollet eventually became an accomplished geological explorer, although before contact with Troost he was not particularly interested in geology. Despite a very close relationship, it seems impossible to demonstrate that Troost actually showed Nicollet how to conduct himself in the field.

Even more vexing is the case of Troost's possible relationship with the English-born George William Featherstonhaugh, who had a long career in American geology. In 1831 and 1832 Featherstonhaugh edited his own geological journal, the *Monthly American Journal of Geology*. In this journal, Featherstonhaugh published dozens of articles on a variety of geological topics. Most were reviews of the literature; none involves extensive original field work.[45]

In the early 1830s Featherstonhaugh was a prominent

person in American geology, perhaps a good deal better known than Troost. Featherstonhaugh was entirely self-trained—a critical reader who attended society meetings, traveled, and corresponded. He was a writer, an editor, and an organizer, but he was not famed for his field skills.

During September and October of 1834 Featherstonhaugh spent two weeks in Nashville, as part of a tour of the South. He visited Troost's home and his classroom. Furthermore, Featherstonhaugh and Troost went on at least two excursions that featured Troost's field interpretation of a fossil assemblage and provided Featherstonhaugh with insight into the stratigraphy and structure of the Nashville area. When Featherstonhaugh published a narrative of his southern tour, he treated Nashville differently from other cities. For most areas, he had cursory comments on geology, but for Nashville he offered detail: two tabulations of the thickness of stratified rocks.[46] Both were based on field measurement and they showed, bed by bed, that one sequence of strata was about 1520 feet thick and the other was precisely 163 feet. This level of field detail seems incompatible with Featherstonhaugh's other publications of the early 1830s.[47] Perhaps he pirated his data from Troost. Or perhaps he participated in the measurement.

After two weeks in Nashville, Featherstonhaugh continued his excursion. He was headed toward the Ozark Mountains and a new phase in his geological career. He had been employed by the federal government to prepare a report, based on original field work, describing the geology of the Ozark region. What influence did Troost have upon the first major field project of the self-educated Featherstonhaugh? Perhaps no influence at all. It is pure speculation, but Featherstonhaugh, Nicollet, Troost, and other pioneer field workers probably interacted. In that interaction, presumably, the more experienced people dominated.

In addition to apprentices and possible apprentices who

later applied their skills professionally, Troost trained at least two people who may never have worked as geologists. One was Lewis Troost, his own son. All biographies of Gerard Troost agree that Lewis graduated from the University of Nashville in 1835 and eventually made civil engineering his career, living near Mobile, Alabama. While he made no known contribution to geology, that he served an apprenticeship with his father is evidenced by one of Gerard Troost's few surviving manuscripts: "Notes Kept during an Excursion in the Months of April and May, 1834."[48]

Thomas Troost, another member of Gerard Troost's household, was the assistant on a second field excursion that is recorded in a surviving manuscript: "Notes Kept during an Excursion in Oct. 1833 in Part of the Western District."[49] For almost fifty years the identity of Thomas Troost has been a puzzle. One of Gerard Troost's biographers, who did not fully utilize manuscript sources, found the name Thomas Troost written in an ancient book in a Nashville library and tried to identify the person involved.[50] He was unsuccessful for an unusual reason: there was no person named Thomas Troost. Just before coming to New Harmony, Gerard Troost had married a widow named Mary O'Reilly, and she had a son named Thomas O'Reilly, who sometimes called himself Thomas Troost. After graduating from the University of Nashville, in 1835, Thomas O'Reilly (Troost) became a civil engineer. While he and Lewis Troost both entered civil engineering at the same time, their careers followed different paths. Apparently, O'Reilly never published a professional report of any kind, but Lewis Troost became the author of many newspaper accounts and propaganda leaflets that were designed to promote the financing of railroads.[51]

Lewis Troost may have surveyed the geology along railroad rights of way, but resulting scientific publications are currently unknown. Since there is still much to learn about nineteenth century scientific literature published in the southern states, it seems best to reserve judgement on the possible geological accomplishments of Lewis Troost.

Conclusions

In the middle 1830s the nature of American geology changed. A blossoming of state-supported surveys heightened public interest and established geology as a professional field. The year 1836 marks a turning point, for it was the year when surveys first proliferated.[52] After 1836 so many state and federal geological surveys offered on-the-job experience to field workers that it is difficult to assess the influence of one individual or one institution on the evolution of American field practices. Before that sudden upsurge of interest the population of geologists was smaller, and it is more feasible to try to identify their cultural roots and the origins of their field methodologies.

After 1836 American field practice was clearly diverse. Contemporary drawings show that some New England field workers did their outdoor work in status-conscious clothing: formal coats, scarves, and top hats.[53] This elegance contrasts sharply with Lesueur's field attire, as shown in Figure 3. Less visible field traditions differed as well. Each field party had to choose what data to collect and what to emphasize. All were inundated with topographic information, stratigraphic data, and random observations on springs, caves, waterfalls, and other geomorphic phenomena. A skilled field worker could spend time collecting or identifying soils, rocks, fossils, and minerals. After that, each party had to strike its own balance between objective description and on-site interpretation. Beyond this, each party had to choose to carry barometers and similar instruments, or to operate without them. The decisions were numerous; so were the skills required.

Before the middle 1830s, most American geologists were

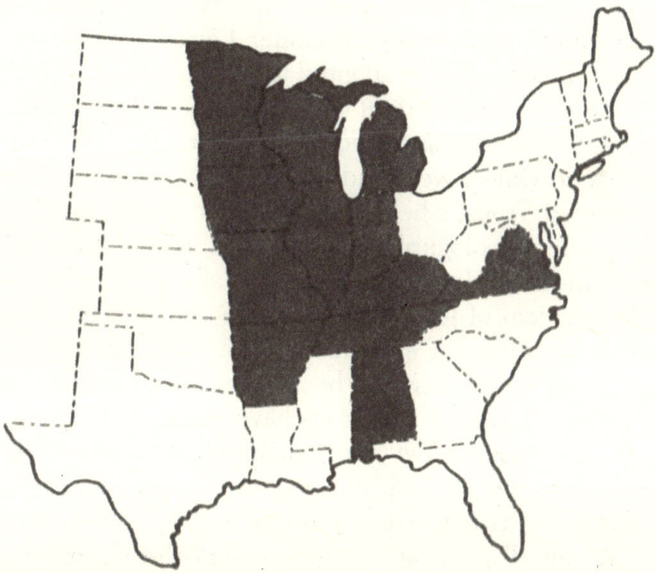

Figure 7.

TROOST'S INFLUENCE. Litton, Owen, and Currey, three of Troost's early apprentices, did significant geological field work in the twelve states indicated in this map. In many of these states, Troost's protégés did the first extensive field work or were the first well-trained observers to prepare detailed studies of economically significant regions. They often studied just a small part of a state, but subsequent geological mapping has generally stressed comparison with earlier work, altering and expanding previous studies. Thus the earliest geological observations have profoundly influenced all later geological interpretations.

largely self-trained, or at least self-trained in field procedures. In continental Europe, long before 1810, an early wave of field workers solved some of the problems of data collection and had evolved a few rudimentary field procedures. The names of these continental pioneers—Von Buch, Cuvier, Desmarest, von Humboldt—were known to all American geologists, but Troost knew them personally.

Through Gerard Troost and his first two field assistants, Benoist Troost and Charles Alexander Lesueur, a European tradition of field work was adapted for use in America. In the early 1830s, before geological surveys became commonplace, Troost's skills were transferred by apprenticeship to Abram Litton, David Dale Owen, and Richard Owen Currey. These three blended Troost's field teachings with other knowledge to make pioneer geological observations in the states shown in Figure 7. In the early 1830s Troost may also have influenced the field practices of Joseph Nicollet, George William Featherstonhaugh, and other early mappers.

The United States has long been viewed as a melting pot, where different cultural traditions have blended. In the evolution of American geology, several European traditions have had great impact on American thought. Because of language, the traditions of Great Britain diffused with special ease, but they were not all pervasive. To the extent that Troost influenced their development, American field traditions in geology are grounded in French culture.

Notes

1. Gary Lee Schoepflin, "Denison Olmsted (1790–1859), Scientist, Teacher, Christian: A Biographical Study of the Connection of Science with Religion in Antebellum America" (Ph.D. diss., Oregon State University, 1977).

2. For a discussion of all early surveys see Mary C. Rabbitt, *Minerals, Land, and Geology for the Common Defence and General Welfare*, vol. 1 (before 1879; Washington, D.C.: U.S. Geological Survey, 1979).

3. Offices held by early European earth scientists are mentioned in Karl A. von Zittle, *History of Geology and Palaeontology to the End of the Nineteenth Century* (London: Walter Scott, 1901).

4. For background on the word *geology* see Frank D. Adams, "Earliest Use of the Term Geology," *Bulletin of the Geological Society of America* 43 (1932):121; and idem, "Further Notes on the Use of the Term Geology," *Bulletin of the Geological Society of America* 44 (1933):831.

5. Sydney Ross, "*Scientist:* The Story of a Word," *Annals of Science* 18 (1962):65–85.

6. Professionalization was also taking place in Europe, where several traditions of geological field work and several traditions of geological draftsmanship vied for prominence. See Roy Porter, "Gentlemen and Geology: The Emergence of a Scientific Career, 1660–1920," *Historical Journal* 21 (1978):809–36; and Martin J. J. Rudwick, "The Emergence of a Visual Language for the Geological Sciences, 1760–1840," *History of Science* 14 (1976):149–95.

7. Biographies of Troost generally state that he prepared a Dutch translation of one of von Humboldt's works, but no English-language biography gives a title for the translation. Professor H. A. M. Snelders of the University of Utrecht kindly provided a formal citation: Gerard Troost, trans., *Natuur-tafereelen, met weterschappelijke ophelderingen*, by Alexander von Humboldt (The Hague, 1808). The work is probably best known through the German edition, *Ansichten der Natur*, which also appeared in 1808.

8. A. Royer, *History and Description of the Royal Museum of Natural History*, translated from the French by M. Deleuze (Paris: Cellet, 1823), p. 106.

9. L. C. Glenn, "Gerard Troost," *American Geologist* 35 (1905):72–94; and H. G. Rooker, "A Sketch of the Life and Works of Dr. Gerard Troost," *Tennessee Historical Magazine*, 2nd ser. 3 (1935 for 1933):3–19.

10. For example, J. van Baren, *Nieuw Nederlandsch biographisch woordenboek*, vol. 5 (Leiden, 1921), cols. 973–74. Reference provided by Dr. H. A. M. Snelders of the University of Utrecht.

11. For example, one of Troost's letters of introduction and his specimen-based information are discussed in a letter, apparently written in the summer of 1810, from a Philadelphia friend to Col. George Gibbs, a noted American mineral collector. While the author's signature is almost unintelligable, it may read C. S. Mece. The letter is in the archives of the Newport Rhode Island Historical Society.

12. Gerard Troost, "Catalogue of a Collection of Minerals formed by G. Troost . . .," vols. 1 and 2, holograph, about 1,400 pages (Kentucky Room, Free Public Library, Louisville, Kentucky).

13. For knowledge of this trip I am indebted to Dr. Ralph Schwarz of Historic New Harmony, Inc. Dr. Schwarz is now building a series of museum exhibits that trace the travels of many early scientists, including Lesueur and Maclure.

14. Biographical data on Benoist Troost are scattered throughout two sources: *The History of Jackson County, Missouri, Containing a History of the County, Its Cities, Towns, etc.* (Kansas City, Missouri: Union Historical Co., 1881); and E. J. Goodwin, *A History of Medicine in Missouri* (St. Louis: W. L. Smith, 1905).

15. Robert M. Hazen, *Bibliography of American-Published Geology: 1669–1850*, Geological Society of America Microform Publication no. 4, 1976. This work was reissued in slightly abbreviated form as Robert M. Hazen and Margaret Hindle Hazen, *American Geological Literature, 1669–1850* (Stroudsburg, Penn.: Dowden, Hutchison, and Ross, 1980).

16. Troost's well-known geological publications are listed by Hazen, *Bibliography*, pp. 913–17. The New Harmony community has been discussed in a thousand biased books. One of the better accounts is Paul Brown, *Twelve Months in New Harmony, Representing a Faithful Account of the Principle Occurrences Which Have Taken Place within That Period, Interspersed with Remarks* (Cincinnati: William Hill Woodward, 1827). Troost's time of arrival is suggested by a manuscript entry in a New Harmony account book that records payment of a tavern bill on January 3, 1826: "New Harmony Community, Petty Cash Journal," 2:216. The amount of the bill, $6.61, was large for the era, indicating that Troost and his family had been there for at least a few days, perhaps a few weeks. They clearly preceded the major influx of intellectuals, the so-called Boat Load of Knowledge. The Boat Load arrived in late January. Mrs. Mary Lou Robson Fleming, of New Harmony, kindly provided data on Troost's arrival. Troost could have lived in New Harmony for many weeks without leaving a written record.

17. E. T. Hamy, *The Travels of the Naturalist Charles A. Lesueur in North America, 1815–1837*, trans. Milton Haber and ed. H. F. Raup (Kent, Ohio: Kent State University Press, 1975). There are many unnumbered plates and drawings by Lesueur. One shows Troost and Lesueur on their excursion.

18. For insight into the antagonisms at New Harmony see Brown, *Twelve Months;* and Arthur E. Bestor, Jr., "Education and Reform at New Harmony, Correspondence of William Maclure and Marie Duclos Fretageot, 1820–1833," *Indiana Historical Collection* 15 (1948):285–417. Caroline Troost, a minor, married Herman B. Taylor in Mount Vernon, Indiana, on February 1, 1827. An annulment was obtained by her father; see Posey County Courthouse, Mount Vernon, Indiana, Record Book C, February

Term, 1827, pp. 142, 420. Mrs. Mary Lou Robson Fleming, of New Harmony, kindly provided these data.

19. James X. Corgan, "Offbeat Pioneers," *Journal of the Tennessee Medical Association* 70 (1977):873–76.

20. For the route they traveled through Tennessee see Hamy, *Travels*, p. 70.

21. Lesueur visited Tennessee on several occasions, and always made sketches of the landscape. About eighty of these drawings still exist, but few have been published. The Tennessee State Museum is now assembling copies. Dr. James Kelly, of the museum, and I hope to prepare an illustrated and annotated catalogue of Lesueur's Tennessee drawings, to be published as a guide to a museum exhibit.

22. There are three biographical sources: Alexander S. Lagsdorf, "The Story of Washington University, 1853–1953," (Manuscript Collection, Missouri Historical Society); Max Goldstein, ed., *One Hundred Years of Medicine and Surgery in Missouri* (St. Louis: St. Louis Star, 1900), pp. 286–87; and G. C. Broadhead, "Abram Litton, M.D.," *Transactions of the St. Louis Academy of Science* 12 (1902):xxiv–xxvii. Lagsdorf's work focuses on later accomplishments; Goldstein's account confuses Minnesota with Mississippi and misspells most names; Broadhead's work errs in most dates but is the best of the trio. Studies by Goldstein and Broadhead were probably based on the memories of a man in his mid-eighties who no longer recalled precise dates and places.

23. Key dates in his career at the University are confirmed by the *Catalogue of the Officers and Graduates of the University of Nashville* (Nashville: A. Nelson, 1850).

24. Broadhead, "Abram Litton," gives the date as July 1840. The chronology of Litton's travels, given later in Broadhead's article, and Litton's dated agricultural report, cited below, make Broadhead's date impossible.

25. A. Litton, "Agricultural Education," *Agriculturist* 2 (1841), p. 139. Litton's report was written in Paris on April 19, 1839. He had been in Europe for many months.

26. A. Litton, "Chemistry," *Agriculturist* 5 (1844), pp. 33–35, 49–50, 65–67.

27. Margaret W. Rossiter, *The Emergence of Agricultural Science: Justus Liebig and the Americans, 1840–1880* (New Haven: Yale University Press, 1975) pp. 184–85.

28. A. Litton to J. Liebig, June 22, 1844, Lane Collection, Missouri Historical Society.

29. For example, "Dr. Troost's Report," *Guardian* 4 (1844):21; and "Professor Litton," *Guardian* 3 (1843):40. Both also mention the field work of Richard O. Currey, discussed below.

30. Rabbitt, *Minerals*, pp. 84–85; George P. Merrill, *The First One Hundred Years of American Geology* (New York: Hafner Publishing Co., 1964), pp. 271–75.

31. Walter B. Hendrickson, *David Dale Owen, Pioneer Geologist of the Middle West* (Indianapolis: Indiana Historical Bureau, 1943), pp. 85–86, 113.

32. The Missouri Historical Society, St. Louis, has an Abram Litton file that includes formal certificates of appointment and copies of laws establishing or renewing geological surveys that employed Litton. Dates given above are well established, even though they may conflict with earlier publications.

33. Hendrickson, *David Dale Owen;* Rabbitt, *Minerals;* Katherine G. Nelson. "Environment for Discovery: The Owen Survey of Wisconsin," *Wisconsin Academy of Science, Arts, and Letters* 64 (1976):173–79; Markes E. Johnson, "Early Explorations of the Silurian System in Iowa," *Proceedings of the Iowa Academy of Science* 84 (1977):150–56; etc.

34. Manuscript records of the Medical College of Ohio are now in the archives of the University of Cincinnati. Linda M. Ellis, of the University Library, kindly verified data. Theses from Owen's era cannot be located.

35. Dr. Michele L. Aldrich, coauthor of a later paper in this volume, called my attention to a letter from James B. Rogers to William B. Rogers that is in the archives of MIT. The letter, written on January 10, 1836, casts light on Troost's stature as a teacher of applied geology. James Riddle, a friend of James Rogers, hoped to find employment with a new state geological survey in Ohio. To strengthen his academic credentials, James Rogers suggested that he spend the summer of 1836 as an apprentice with Troost in Tennessee. Riddle did not follow his friend's advice. If he had, he would have joined David Dale Owen in learning field procedures. Owen, of course, was another friend of the Rogers family.

36. Johnson, "Geology in American Education: 1825–1860," *Bulletin of the Geological Society of America* 88 (1977):1192–98.

37. Samuel J. Platt and Mary L. Ogden, *Medical Men and Institutions of Knox County, Tennessee, 1789–1957*. (Knoxville: S. B. Newman Printing Co., 1969), pp. 250–51; L. C. Glenn, "The Growth of Our Knowledge of Tennessee Geology," *Tennessee Geological Survey, Resources of Tennessee* 2 (1912):167–219; Paul F. Eve, "Brief Biographical Sketches of the Professional Dead of Middle Tennessee," *Transactions of the Medical Society of the State of Tennessee* (1876), pp. 78–90.

38. James X. Corgan, "Tennessee's Early Technical and Scientific Journals, 1825–1861," *Journal of the Tennessee Academy of Science* 52 (1977):23–26. In this report, Currey's *Southern Agriculturist* is identified as a planned journal that may never have been published, but it is now clear

that at least one issue appeared. The first number, issued in January 1851, was reviewed the following month: "Southern Agriculturist," *Southern Lady's Companion* 4 (1851):262. On Currey's teaching, see James X. Corgan, "Some Firsts? in the Colleges of Tennessee," *Journal of the Tennessee Academy of Science* 55 (1980):86–91. For a partial list of the societies he organized see James X. Corgan, "A Preliminary Catalog of the Scientific and Technical Societies of Antebellum East Tennessee," *East Tennessee Historical Society's Publications* 48 (1980 for 1976):21–28.

39. Fred B. Weaver, "Medical Care in a Confederate Prison," *North Carolina Medical Journal* 25 (1964):206–9.

40. Richard O. Currey, *A Sketch of the Geology of Tennessee* (Knoxville: Haskell, 1857); and idem, "An Essay on Iron for the Degree of Doctor of Medicine in the University of Pennsylvania," January 1840, manuscript collection, University of Pennsylvania.

41. Richard O. Currey, *A Geological Visit to the Virginia Copper Region* (Knoxville: Beckett, Haws and Co., 1859).

42. Richard O. Currey, "The Paper on the Copper and Other Resources of the Floyd-Carroll-Grayson Plateau," *The Virginias* 1 (1880): 60, 62–64, 69–71, 74–77, 80–81, 95, with illustrations after p. 60.

43. Richard O. Currey, "Geology of Benton County, Ala.," *Southern Journal of Medical and Physical Science* 2 (1854): 195–202.

44. For example, a long list of reports on Tennessee is cited by Charles W. Wilson, Jr., *Annotated Bibliography of the Geology of Tennessee . . .*, Tennessee Division of Geology Bulletin no. 59, 1953.

45. Featherstonhaugh is treated in all standard biographical dictionaries. His *Monthly Journal* has recently been republished, with an introduction by George W. White: G. W. Featherstonhaugh, ed., *The Monthly American Journal of Geology and Natural Science* (New York: Hafner Press), vol. 1, *1831–1832*.

46. George W. Featherstonhaugh, *Excursion through the Slave States, from Washington on the Potomac to the Frontiers of Mexico; with Sketches of Popular Manners and Geological Notices* (New York: Harper and Brothers, 1844).

47. Cited in Hazen, *Bibliography*.

48. This manuscript is part of a large, complexly paginated notebook that is indexed as Troost, "Diary" in the Tennessee State Archives, Nashville.

49. Troost, "Diary."

50. Rooker, "Sketch." A still unsuccessful search for a first job for Thomas O'Reilly (Troost) was discussed in a letter from Gerard Troost to Joseph Nicollet, dated October 1835. The letter is in the Miscellaneous Papers of the New York Public Library.

51. He wrote several railroad reports, the best known of which is Lewis

Troost, *Report on the Preliminary Surveys of the Mobile and New Orleans Railroad* (Mobile, 1854). His known reports were promotional literature, selling the concept of a new service.

52. Rabbitt, *Minerals*, discusses each survey.

53. Rabbitt, *Minerals*, figure on p. 42.

Mineral Fertilizers in Southern Agriculture

Richard C. Sheridan

Much of the farmland in Virginia and the other original thirteen states had become worn out and eroded by 1800 or even earlier. One reason for the rapid exhaustion of this land was the failure of farmers to rotate crops. Another cause was neglect in replacing the organic material that constituted the chief source of nutrients in the soil. Little animal manure was available because the livestock generally ran at large, rather than being kept in pens or barns. Furthermore, in the early days land was plentiful, and when a field ceased to yield a good crop, it was an easy matter to clear new land for cultivation. Soon many families began to move west or south to new, virgin lands. As a result, land values in the older counties decreased and the economy became stagnant.

A few American farmers began to use calcareous materials in the later part of the eighteenth century. As early as 1748 limestone was used on cornfields near Brunswick, New Jersey. In 1775 many Pennsylvania farmers were calcining limestone for use on their fields.[1] Lime also was used on a small scale in Virginia before the Revolution.

In the period following the close of the Revolutionary War, George Washington experimented with fertilizers such as marsh mud, marl, plaster of Paris, ground limestone, and different green manures. Washington kept many of his slaves employed in hauling mud from the Potomac to his fields.[2]

An important development shortly after the war was the introduction of gypsum as an inorganic fertilizer. Gypsum, a calcium sulfate mineral, also was known as plaster of Paris and land plaster. The first gypsum used in America for fertilization came from Paris, and was imported and used by Benjamin Franklin. About 1786 to 1790 the use of gypsum spread rapidly in Virginia. John Binns, of Loudoun County, Virginia, published a small book in 1803 in which he described how he had used gypsum to convert his "entirely worn-out" farm into a "rich and fruitful one." His booklet was widely read, and Thomas Jefferson gave him credit for restoring Loudoun County to prosperity and stopping emigration to the lower South.[3] Binns' favorite method of applying land plaster was to roll the moistened seeds in it before sowing. He supplemented this treatment with applications placed on the hill or sown on the crop in the early spring, with a total application of about two bushels or less to the acre per year.[4]

Numerous experiments with gypsum were reported also by John Taylor, of Caroline County, Virginia, in 1809.[5] Gypsum soon came to be used in connection with clover and crop rotation by the better farmers in northern Virginia, western Maryland, and the upper valley of Virginia. The combination of clover, gypsum, and crop rotation worked a great transformation in the agriculture of those areas. The clover was plowed under, thereby supplying both nitrogen and organic matter to the soil so that other crops could be advantageously grown. However, the use of gypsum never caught on with farmers in the coastal plain of Virginia and the Carolinas, except on a few plantations along the James River. Some of these farmers believed that gypsum was unsuited to soils located near salt water. Others probably received disappointing results by applying it directly to crops instead of in combination with clover.[6]

Most of the gypsum used in southern seaboard states was imported from the Canadian maritime provinces of Nova Scotia and New Brunswick. About 30,000 tons were imported into the

United States in 1800, but the annual import soon climbed to 100,000 tons. Except during the War of 1812, gypsum usually sold at retail prices of $4.50 to $5.00 per ton.[7] Gypsum was delivered in barrels and was frequently hard and lumpy, so that it had to be pulverized before spreading. Pulverizing was sometimes done by water-powered hammers at a gristmill.[8]

Gypsum supplies two essential elements to plants—calcium and sulfur—and it aids in the accumulation of nitrogen by its effect on nitrogen-fixing legumes, but it is not a complete fertilizer. Where other nutrients are deficient, such as phosphorus or potassium, gypsum gives no returns. Under these conditions, legume production soon depletes the phosphorus and potassium, leaving such soils even more impoverished. Thus, gypsum and legumes provided only a temporary relief to soil infertility. Failure to supply the major nutrients was the main cause of the demise of gypsum; the invention of commercial fertilizers containing nitrogen, phosphorus, and potassium finally provided the nutrients that were really needed.[9]

The use of lime was reintroduced into lower Virginia about 1794 by Fielding Lewis, and experiments with it occurred in South Carolina as early as 1800. Nevertheless, for many years lime did not generate the great interest that had arisen in connection with gypsum. Some planters in Maryland and Delaware made a practice of burning oyster shells and then applying the resulting lime to their fields. Later, lime was produced from the extensive limestone beds in the upper valley of Virginia and transported to middle Virginia, which did not possess abundant marl deposits.[10] About one cord of wood was consumed for each fifty bushels of lime produced. In the 1830s some Maryland farmers purchased lime from Pennsylvania kilns, where it was produced by burning limestone with anthracite coal. The lime sold at the kilns for ten cents per bushel, but transportation charges easily doubled or tripled the cost to the farmer.[11] In 1850 the annual national consumption of burnt lime was 280,000 tons; by 1860 it had increased to 310,000 tons.[12] The recommended rate of application to the soil was thirty to fifty bushels per acre.[13]

Edmund Ruffin was the most influential leader of southern agriculture in the antebellum period.[14] Born on a Virginia plantation in 1794, Ruffin inherited the property when he was a young man and tried to improve the soil by using the practices recommended by John Taylor. He was unsuccessful until he read Sir Humphrey Davy's *Elements of Agricultural Chemistry*, which encouraged him to test the soil. He soon concluded that soil acidity was responsible for his poor yields and that carbonate of lime was the remedy. Before long, Ruffin found that marl improved the soil and increased his yield of crops. Marl is a soft, chalky deposit found in the bottom of former lakes and streams, consisting mainly of clay and amorphous calcium carbonate. It also is found in the form of partly decomposed fossil shells. Ruffin did not claim to be the original discoverer; three or four Virginia farmers were using marl shortly before 1820, and John Singleton, of Talbot County, Maryland, also had tried marl in 1805. These farmers, however, had no definite objective in view other than general improvement of the land.[15]

In 1818 Ruffin excavated marl containing about 33 percent calcium carbonate from beds on his own farm and applied it at the rate of 125 to 200 bushels to the acre. From the land thus treated he obtained an increase of 40 percent over the crop on similar, untreated land. In 1821 he presented the results of his experiments at a meeting of the Agricultural Society of Prince George County, and published the paper in the *American Farmer*. For the next thirty to thirty-five years calcareous fertilizers were the central concern of Ruffin's life, although in later life he became deeply involved in the slave issue and abandoned his efforts at soil improvement. For many years he conducted carefully planned experiments. He analyzed soils, searched for marls and determined their lime contents, and tested the effects of marl with different animal and vegetable manures.[16]

Figure 1.
EDMUND RUFFIN. A rabid secessionist, Edmund Ruffin joined South Carolina's Palmetto Guard and set off the first artillery shell fired at Fort Sumter. He also participated in the Battle of Bull Run, but his age (sixty-seven) precluded continued military service. In retaliation against his part in the war, federal gunboats frequently took pot shots at his home on the James River, and Union troops pillaged his estate and vandalized his library.

Ruffin's first paper, somewhat expanded, was published in book form as *An Essay on Calcareous Manures* in 1832.[17] New editions of the essay were published at frequent intervals, with the fifth appearing in 1852. During this twenty-year period the book grew from its original 242 pages to 493 pages in the final edition. Most of the enlargement was due to the addition of directions for finding and analyzing marl beds. Ruffin also gave information on the cost of marling land and reviewed the progress of marling in Virginia after 1818. Ruffin taught that marl neutralized soil acidity, improved the texture of soils, promoted early maturity of crops, and supplied a missing plant nutrient. He was the editor of a monthly magazine, the *Farmer's Register,* for ten years and wrote many of the articles in it. Marl and its use were discussed repeatedly.

As a result of Ruffin's crusade, many of the more progressive farmers and planters began to use marl, especially in lower Virginia, Delaware, and Maryland, where beds were readily available. The productivity and value of these farms was increased three or four times by the use of marl and clover. Ruffin's work stimulated interest in marl in other states, and in 1842 he was asked to make a survey of the lime deposits in South Carolina. Some South Carolina planters were induced to use marl, but Ruffin admitted that his preaching had little effect in that state. In eastern North Carolina marl had been used as early as 1818, and in 1825 Professor Denison Olmsted had called attention to the extensive marl deposits of the state. As late as 1847 there was said to be little use made of it, but much enthusiasm for marling developed in the late fifties.[18] The consumption of marl in the United States was 500,000 tons in 1850 and about 600,000 tons in 1860.[19]

While gypsum, marl, and lime were widely used in the states from Maryland to South Carolina, they were not in common use throughout the South. In Alabama and Mississippi, for example, fertilizers were seldom used before the Civil War, probably because the fields were still fairly fertile. However, some fertilizer materials, including lime, marble dust, and gypsum, were sold to farmers at Mobile and probably at other Gulf ports after 1857.[20]

While a few mineral fertilizers became mainstays of scientific agriculture in the antebellum South, there also was a lot of experimentation with less widely used materials. Ordinary salt, saltpeter, barium sulfate, and greensand marl were sometimes applied as fertilizers. In 1850, 7,860 tons of fertilizer-grade saltpeter (potassium nitrate) was imported from Chile and India; by 1860 the importation of this fertilizer exceeded 10,000 tons per year.[21] Experiments by a Dr. Barton of Rockbridge County, Virginia, showed that an application of ground barium sulfate gave favorable results for five years.[22] Greensand marl found in New Jersey and Delaware contained significant amounts of potash and small amounts of phosphate.[23] Ruffin used greensand on his plantation, but he did not seem to have recognized the value of these nutrients and based his opinion of it only on the calcium carbonate content.[24]

Calcium phosphate is the chief constituent of bones, and crushed bones were used as a fertilizer as early as 1825 in Maryland. The use of bonemeal gradually increased and amounted to 10,000 tons in 1860.[25] In 1837 F. S. Holmes and other geology students found a number of water-worn nodules or rocks scattered over a field near the Ashley river in South Carolina. Mingled with the rocks were many interesting and well-preserved shells, teeth, and bones. When Edmund Ruffin arrived to make a survey for marl and lime in 1843, Holmes submitted specimens of the nodular rocks for his examination. However, as the rocks contained little carbonate of lime, the nodules were tossed aside and considered useless as a fertilizer material. At about the same time John S. Brisbane of St. Andrews had a large quantity of the rocks crushed, intending to use it as a fertilizer, but he was advised that it would produce no beneficial effect.[26] Professor Charles U. Shepard was the first to indicate that the rocks contained calcium phosphate. In 1859 Shepard explained the nature of the mineral in a lecture before the medical society of South Carolina, and in 1860 he advised the use of the Ashley phosphates in lieu of bones. The outbreak of the Civil War, however, put a stop to fertilizer operations, and it was not until 1867 that Professor Holmes and Dr. N. A. Pratt realized the composition, extent, and value of the rocks.[27]

Commercial exploitation of the South Carolina phosphate deposits began in 1868, and a thriving fertilizer industry was quickly established. This was the beginning of an industry that today produces about fifty million tons of phosphate per year— not only for southern agriculture, but also for use throughout the nation and the world.

Names and Composition of Mineral Fertilizers Used in the Antebellum South

Common name	Principal constituent	Chemical formula
Gypsum (also known as land plaster, plaster of Paris, and sulfate of lime)	Calcium sulfate	$CaSO_4 \cdot 2H_2O$
Limestone	Calcium carbonate	$CaCO_3$
Marble dust	Calcium carbonate	$CaCO_3$
Lime (also known as burnt, caustic, or quick lime)	Calcium oxide	CaO
Slaked lime	Calcium hydroxide	$Ca(OH)_2$
Salt	Sodium chloride	$NaCl$
Saltpeter	Potassium nitrate	KNO_3
Sulfate of barytes	Barium sulfate	$BaSO_4$
Marl	Clay and amorphous calcium carbonate	—
Greensand	Potassium iron silicate	—

Notes

1. Percy W. Bidwell and John I. Falconer, *History of Agriculture in the Northern United States, 1620–1860* (Clifton, N.J.: Augustus M. Kelley Publishers, 1973), p. 88.

2. Lewis C. Gray, *History of Agriculture in the Southern United States to 1860*, vol. 2 (Clifton, N.J.: Augustus M. Kelley Publishers, 1973), pp. 612, 803.

3. Ibid., p. 803.

4. William Crocker, *The History of Agricultural Gypsum* (Chicago: The Gypsum Industries Association, 1922), p. 9.

5. Wayne D. Rasmussen, *Agriculture in the United States: A Documentary History* (New York: Random House, 1975), p. 403.

6. Gray, *History*, pp. 803–4.

7. Gerald S. Graham, "The Gypsum Trade of the Maritime Provinces," *Agricultural History* 12 (1938):209–23.

8. G. M. Herndon, "Agricultural Reform in Antebellum Virginia: William Galt, Jr., A Case Study," *Agricultural History* 52 (1978):398.

9. Crocker, *Agricultural Gypsum*, p. 15.

10. Gray, *History*, p. 804.

11. Bidwell and Falconer, *History*, p. 233.

12. A. L. Mehring, "Fertilizers," in *Handbook of Food and Agriculture*, ed. Fred C. Blanck (New York: Reinhold Publishing Corporation, 1955), p. 93.

13. Rasmussen, *Documentary History*, p. 410.

14. Gray, *History*, p. 804; Angus McDonald, *Early American Soil Conservationists*, U.S. Department of Agriculture, Miscellaneous Publication no. 449 (1941), pp. 42–54; W. P. Cutter, "A Pioneer in Agricultural Science," *Yearbook of the United States Department of Agriculture, 1895* (Washington: U.S. Government Printing Office, 1896), pp. 493–502; Aaron J. Ihde, "Edmund Ruffin, Soil Chemist of the Old South," *Journal of Chemical Education* 29 (1952):407–14.

15. Gray, *History*, p. 780.

16. Ibid., p. 781.

17. Edmund Ruffin, *An Essay on Calcareous Manures* (Cambridge, Mass.: Harvard University Press, 1961).

18. Gray, *History*, pp. 804–5.

19. Mehring, "Fertilizers," p. 93.

20. *Mobile* (Ala.) *Daily Register*, November 1, 1857; November 5, 1857; January 8, 1858.

21. A. L. Mehring, J. Richard Adams, and K. D. Jacob, *Statistics on*

Fertilizers and Liming Materials in the United States, Statistical Bulletin no. 191, U.S. Department of Agriculture, 1957, p. 30.

22. Thomas G. Clemson, "Fertilizers," in *Report of the Commissioner of Patents for the Year 1859, Agriculture* (Washington: George W. Bowman, Printer, 1860), pp. 163–65.

23. Ibid., p. 155.

24. Cutter, "Pioneer," p. 499.

25. K. D. Jacob, "Predecessors of Superphosphate," in *Superphosphate: Its History, Chemistry, and Manufacture* (Washington: U.S. Government Printing Office, 1964), pp. 10–11.

26. Francis S. Holmes, *Phosphate Rocks of South Carolina and the Great Carolina Marl Bed* (Charleston: Holmes' Book House, 1870), pp. 56–57, 60.

27. William H. Waggaman, "A Report on the Phosphate Fields of South Carolina," *Commercial Fertilizers* 7 (1913):33.

William Barton Rogers and the Virginia Geological Survey, 1835–1842

Michele L. Aldrich and Alan E. Leviton

In March 1835 the Virginia Legislature authorized a one-year geological reconnaissance of the state under the administration of the Board of Public Works, which oversaw internal improvements such as roads, canals, and railroads. Less than a month later the newly appointed geologist, William Barton Rogers, then professor of natural history at William and Mary College, began the field work to which he was to devote virtually all of his time for the next six months when free of teaching responsibilities. During this brief period, he ran one lengthy cross-section (geologic profile) of about 500 miles through geologically and topographically complex country (Figure 1), much of it wilderness and difficult to traverse, constructed several transverse and local sections to fill in details, and in general framed the questions that were to determine his direction as Virginia state geologist for the next six years.[1] It also led him to decide which regions of the state to emphasize, and it suggested geological problems to which he should devote his time. And, although Rogers' reconnaissance of 1835 was a remarkable achievement for its time—as it would be remarkable for any period in the history of American geology—regrettably the achievement has been obscured by his failure to write a final survey report. However, in a sense, he produced his final report first, in his

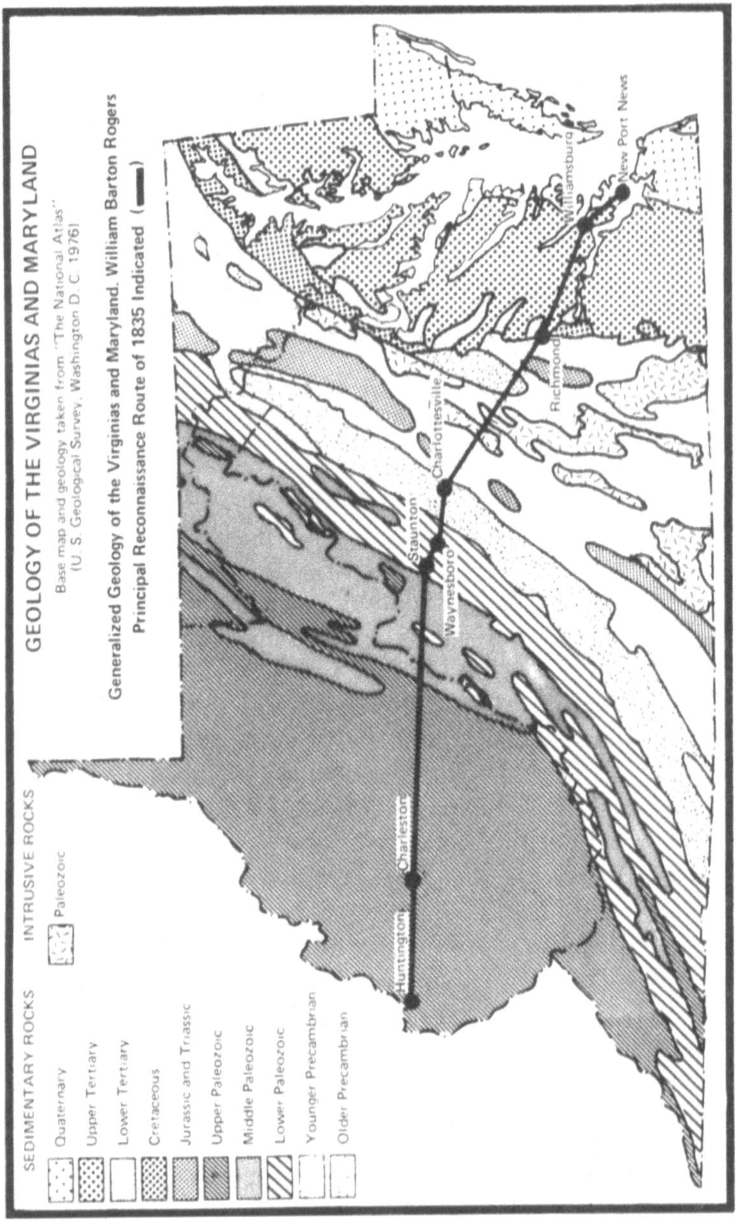
GEOLOGY OF THE VIRGINIAS AND MARYLAND
Base map and geology taken from "The National Atlas"
(U. S. Geological Survey, Washington D. C. 1976)
Generalized Geology of the Virginias and Maryland. William Barton Rogers
Principal Reconnaissance Route of 1835 Indicated
SEDIMENTARY ROCKS
Quaternary
Upper Tertiary
Lower Tertiary
Cretaceous
Jurassic and Triassic
Upper Paleozoic
Middle Paleozoic
Lower Paleozoic
Younger Precambrian
Older Precambrian
INTRUSIVE ROCKS
Paleozoic
Huntington
Charleston
Staunton
Waynesboro
Charlottesville
Richmond
Williamsburg
New Port News

Figure 1.

masterful 1836 summary of this reconnaissance of a region nearly 65,000 miles square.

Born on December 7, 1804, in Philadelphia, William Barton Rogers was the second of four sons of Hannah and Patrick Rogers, the latter a chemist and physician who taught at William and Mary during the boy's childhood. William was educated at that school and later taught both chemistry and natural history there from 1828 until 1835, when he took a job at the University of Virginia in Charlottesville.[2] Of his three brothers, who were also scientists, Henry Darwin Rogers, born in 1808, had the greatest influence on William's thought and work. In 1832 Henry went to England, attracted by the radical educational and social ideas of Robert Owen. There he quickly fell in with the English geologists of the time; he visited H. T. De La Beche, then doing the survey of Cornwall, in 1833, and met the theoretical lions of British geology, ranging in scientific philosophy from Lyell to Murchison.[3]

On his return to the United States, Henry infected his elder brother with enthusiasm for geology. But the idea of a geological survey such as De La Beche's in Cornwall was no novelty in America; both North Carolina and South Carolina had funded modest surveys in the 1820s, the former producing several slim but intriguing reports associated with internal improvements (which would be true of the Virginia survey a decade later). Further, Massachusetts had just completed a survey done by Edward Hitchcock, who had issued a massive report in 1833, and Maryland had a survey under way, as did Tennessee. Finally, New Jersey and Connecticut were to start their surveys in 1835, the same year as William Barton Rogers' reconnaissance, New Jersey choosing William's younger brother Henry as its state geologist.[4]

Agitation for a survey of Virginia had come from the North. Peter Browne, the secretary of the Geological Society of Pennsylvania, had suggested in a letter to Virginia's Governor John Floyd, in September 1833, that a survey be undertaken

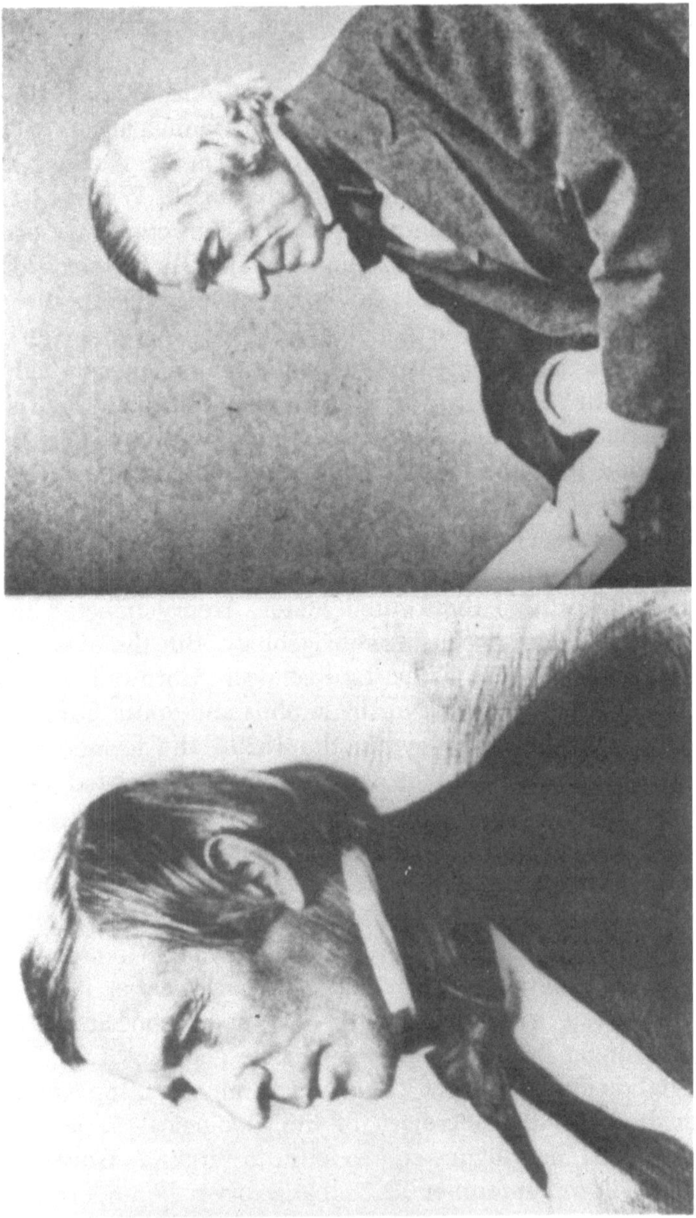

Figure 2.
Portraits of Henry Darwin Rogers (left) and William Barton Rogers (right).

in that state.[5] In December of that year Floyd endorsed the idea in his message to the Virginia legislature. He associated the survey with the building of roads and the improvement of agriculture, stating further that the survey would bring mineral wealth to the state to supplement its other resources. Commerce and capitalism were comfortable words for him.[6] Clearly Floyd was not a typical southern governor with a fixation on farming as the only decent pursuit for the citizens of his commonwealth. And in the next legislative session, Floyd's idea of a survey was reinforced by petitions from Morgan, Shenandoah, and Frederick counties (all in the North Mountain region, to which Rogers was to devote considerable attention).[7] The request of these counties led to the appointment of a committee that met with Rogers and to which he lectured for an hour and a half. The committee was so impressed that its members asked Rogers to speak to the legislature as a whole. William, in a letter to his brother Henry, reported on that event:

> With but a few hours' warning and without a note, and without even casting a thought on how I was to address them and with only one illustration (a magnified section), I marched into the hall of delegates yesterday evening at half past seven. At least three hundred persons had already appeared. . . . It might well have daunted a stouter heart than mine. But a scarcely momentary tremor gave way to the conscious feeling of the importance and dignity of the occasion, and I stood forth boldly and advocated, I think powerfully, the cause of geology, developing a few of its most important truths, and displaying the benefits which it profferred to Virginia. I was listened to with a riveted and deep attention, which satisfied me of the interest which I excited; and without once halting or stammering or becoming confused I went on for upwards of an hour, and when I closed loud words of approbation followed me.[8]

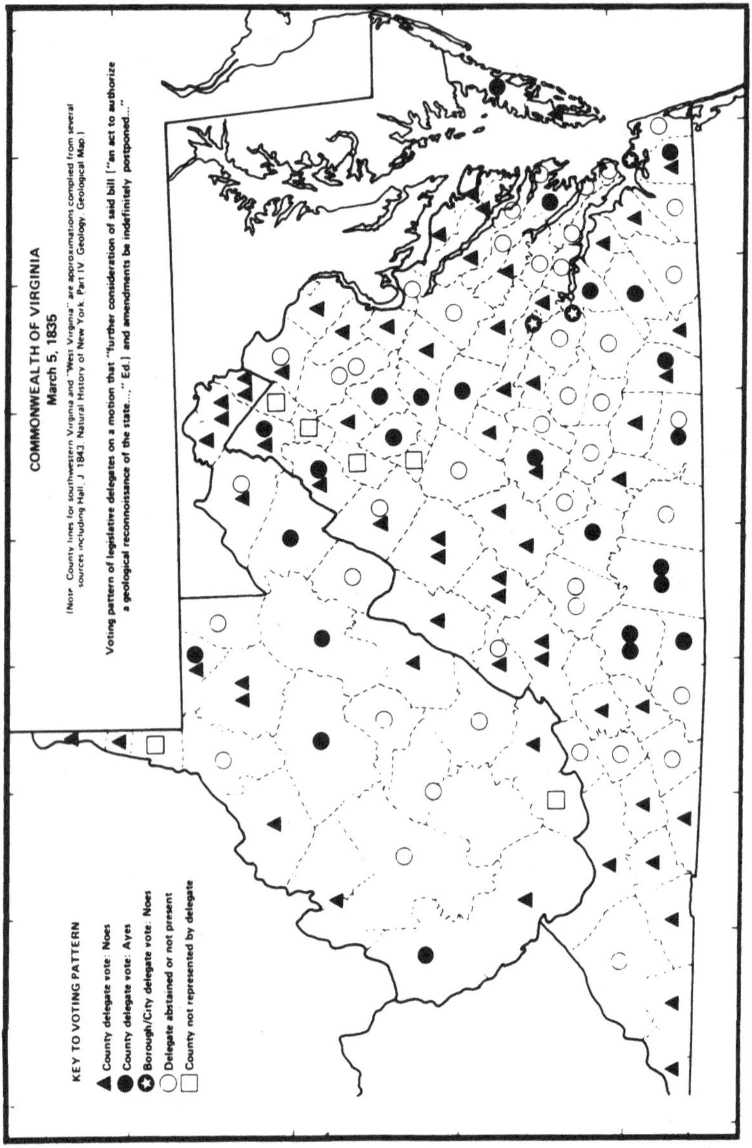

Figure 3.
Geographical distribution of delegates to the Virginia State Legislature voting for (noes) and against (ayes) the geological reconnaissance, 1835.

The legislature passed the bill in March 1835. The House of Delegates split on the measure 69 votes to 26, and as shown in Figure 3, no strong geographical alignments explain the distribution of favorable or hostile votes.[9]

The bill that emerged from the legislature authorized the Virginia Board of Public Works to commission a general reconnaissance, a chemical analysis of soils, minerals and spring water, and a plan for a detailed survey of the state.[10] Rogers had already been at work on the geology of the state for more than a year. He had prepared papers on the marl formations and on the Tertiary fossils, which were published in the *Farmers Register* and the *Transactions of the American Philosophical Society*. In these he described several new species of shells, announced that green sand occurred as grains throughout the marls, and reported his discovery of an Eocene formation in the state.[11] On a geological map of Virginia, the Eocene is not overwhelming (see Fig. 1, undifferentiated Lower Tertiary) and locating it was not easy. Rogers determined ages for Tertiary formations using Lyell's recently published method of computing percentages of modern and extinct shells, which marked Rogers as a relatively sophisticated geologist for that time.[12]

Rogers' pre-1835 work does not preempt what he accomplished in the state-financed reconnaissance of 1835. Working with only occasional assistance from his brother Henry, William Rogers made his first attempt to synthesize the geology of this complex state. For stratigraphic terms, he used names based on texture, color, lithology, and mineral contents. He did not employ names based on type locality, a practice later made famous by the New York survey; nor did he start in this report the practice of numbering formations, which Henry developed a few years later and advocated despite its widespread rejection by other geologists.[13] He refused to correlate the formations of Virginia with those of Europe, saying such interpretation was premature; he felt it was better to de-

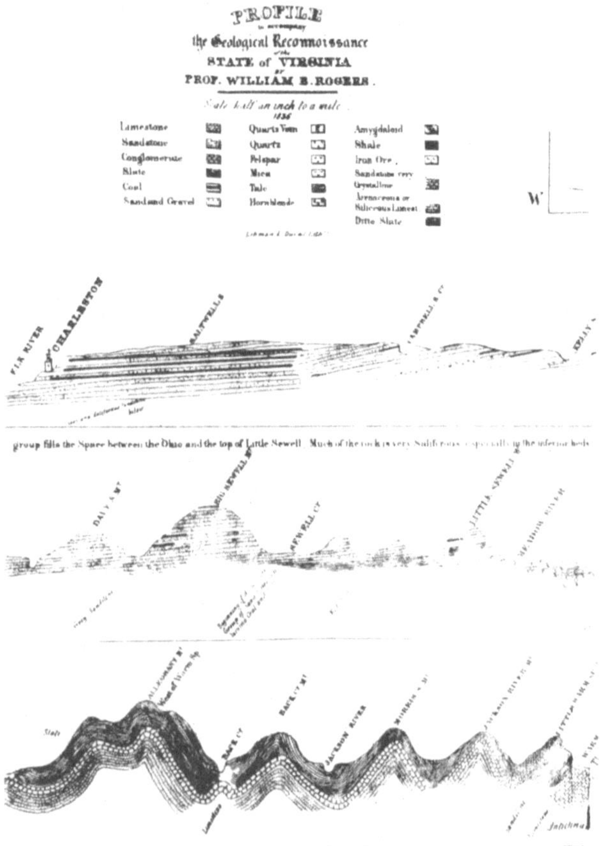

Figure 4.
Part of the profile from Rogers' report of the 1835 reconnaissance.

scribe, not analyze, at this early stage of work. He continued to use fossils to date formations, at least to assign them general ages, and he also used the conditions of the fossils' deposition as a clue to the geological setting at the time.

Rogers reported on a few topics in some detail in the reconnaissance—especially mineral springs, the gold region, and the marls. His study of structure concentrated on folding, not faulting, as the profile to the report illustrates (Fig. 4). His discussion of the origin of the gold-bearing quartz veins, however, is particularly instructive, for it demonstrates Rogers' infinite patience with details, though he did not lose sight of a broader overview. He communicated useful information in a simple, lucid style. Step by step the reader is led through the complexities of the observations and argument favoring the view that the veins are not of the same age as the enclosing rocks, and that they were forcibly injected into the latter by igneous agencies from beneath and at a later date. Rogers' distinguished contemporary, Benjamin Silliman, did not agree with these views.[14] Writing nearly a hundred years later, John Lonsdale observed that Rogers was quite right regarding the relative ages of the veins, and that his interpretation of their formation is "essentially the view held today."[15]

Rogers' report of 1835 closed with a proposed plan for a detailed geological survey of the state. He recommended that the state geologists produce a descriptive geology for all parts of the state; construct a geologic map, using the existing topographical map of the state, with errors in the base corrected as geological work progressed; and assemble a cabinet—museum in our time—of mineral specimens described in the report to demonstrate their practical use. Rogers advised that state scientists investigate several districts simultaneously, progressing across sections and not following the lines of formation across country. He also suggested that barometrical observations be made both of the heights of the mountains and the depths of the gorges and passes, over which roads might be carried, and that observations be made on the drainages or water power of the state.

The bill creating the detailed survey passed early in 1836.[16] To Rogers' plan, the legislature added provisions for sets of rocks to be prepared for colleges in the state, for the publication of annual and final reports, and for a glossary of scientific terms. The statute was unusual for state surveys at the time because it did not specify how long the survey would be conducted.[17] Theoretically, it could stretch out indefinitely.

In conducting his survey, Rogers took the advice of his brother Henry, who was then supervising both the Pennsylvania and New Jersey surveys, and who depended heavily on assistants to do the field work: "Under no circumstances ought the State to look to us for detailed work; that is to come from the assistants. Be not teased from what you hear, but quell the impatience we have awakened, by pleading the importance of maturing plans and getting organized with fitly instructed assistants ere we go minutely to work."[18]

Table 1
Assistant Geologists on the Virginia Survey*

Year	Assistant	Assignment
1836	Henry Darwin Rogers	Virginia-Pennsylvania border.
1837	Charles B. Hayden	Tidewater region along the James River, and northern section of the region between the Blue Ridge and the bituminous coal fields.
	James B. Rogers	Middle section of the area between the Blue Ridge and the bituminous coal fields.
	William Aikin	Southern section of the region between the Blue Ridge and the bituminous coal fields.
1838	William Aikin	Same assignment as 1837.
	Charles B. Hayden and Israel Slade	Northern section of the region between the Blue Ridge and the bituminous coal fields.
	George W. Boyd	Western flank of the Blue Ridge.
	James B. Rogers and George W. Boyd	Middle section of the region between the Blue Ridge and the bituminous coal fields.

Table 1—*Continued*

Year	Assistant	Assignment
1839	George W. Boyd	Middle section of the Piedmont region.
	Charles B. Hayden	Tidewater region between the James River and North Carolina; and Pendleton, Preston, Hampshire, Hardy, Randolph, and Harrison counties.
	James B. Rogers	Mill Mt. and neighboring ridges to the east; lower end of Falling Spring Valley; the region between the Yew and Flat Top mountains; and the Kanawha River Valley.
	Israel Slade	Southern section of the region between the Blue Ridge and the bituminous coal fields.
	Caleb Briggs	Area between the Ohio River and Pennsylvania border, and Monogalia and Harrison counties.
1840	George W. Boyd	Piedmont, between the Potomac and James rivers.
	Caleb Briggs	Ohio River to Kentucky line; Little Kanawha Valley; and the Monongahela Valley.
	Israel Slade	Same assignment as 1839.
	Samuel Lewis	Piedmont region between the James River and North Carolina; and the coal basins of Goochland, Henrico, Powhatan, and Chesterfield counties.
	Thomas Ridgway	Region between the Kanawha River Valley and the Kentucky line.
1841	Caleb Briggs, Thomas Ridgway, and M. Wells	Coal region of the western part of the state.
	Israel Slade	Portions of the Appalachian Mountains.
	James B. Rogers	Piedmont region and upper section of tidewater region (Mt. Vernon to Rappahannock River).

*Source: Rogers, *Geology of the Virginias*, pp. 134, 149-50, 156, 192, 247-48, 413-15, 540.

As each assistant came on (see Table 1), and in order to acquaint him with the geology, William Rogers sent him to do cross-sections (or geologic profiles) for the region to which he would be assigned. Based on his reconnaissance of 1835, Rogers had a good idea of what questions needed to be addressed, and he provided a framework and even hypotheses for his assistants. The assistants, in turn, depended on local residents with an interest in geology to guide them to the outcrops, which were rare in many parts of the state. Often Rogers would visit the assistants to review particularly important sites. But, unfortunately, there were several problems with this way of conducting a survey. For instance, Rogers no sooner trained an assistant than he lost him to employment elsewhere. And, although members of his corps often had experience on other surveys, such as those of New York or Ohio, still some of them were not very good geologists. Furthermore, their letters (field reports) frequently contained a mass of local detail that Rogers alone had to fashion into a coherent report at the end of each field season.[19]

One exception was Dr. George Boyd, a veteran of the New York corps, whose letters reveal his keen eye for stratigraphy. Boyd seems to have written his notes in a form which Rogers could readily use in his reports.[20] One of the saddest episodes of the survey was Boyd's death in 1840 from lung ailments, probably related to overwork in the field. Rogers attended Boyd through his last painful illness, and the correspondence about this tragedy makes very moving reading.[21]

These were some of the personal problems of Rogers' survey. The intellectual drawbacks of a heavy dependence on assistants were even more serious. Toward the end of the Virginia survey, William Rogers realized he had made a mistake in adopting Henry Rogers' model of administration. He complained that "assistants omit things which a proper knowledge of the true objects and mode of research would have suggested as important." Except for his brother James' work on geological structure, William's own brief forays into the field were more fruitful "than months of the comparatively blind labor of . . . assistants."[22]

In fairness to the assistants, it should be said that doing field work in Virginia in the 1830s was no picnic. Charles B. Hayden wrote an especially graphic account of his work in the mountains in 1838. "The exposures are rare and unsatisfactory, the country wild and inaccessible . . . except by foot & inhabited by the lowest order of peasantry poor in every thing but bed-bugs fleas & other pestiferous insects of whose wealth in which we still retain lamentable proofs."[23] He later expanded on this theme, saying of another county that "the bedbugs are perfect snapping turtles."[24] The lack of outcrops and the difficult mountainous terrain were special trials. The work of the nineteenth-century Virginia survey was akin to modern investigations in the thinly populated, densely forested, geologically unexplored tropics. Historians of the transportation movement in America commented that railroad engineers hit the most difficult challenges at the beginning of their work: crossing the Appalachians. After that, the Midwest seemed like child's play, and called for different talents, mainly managerial.[25] Similarly, American geologists found themselves faced with an exceedingly tough terrain in the early years of their discipline, and it is little short of miraculous that they accomplished as much as they did.

Doing geology in early Virginia was not uniformly difficult, however. The tidewater investigators found their life pleasant indeed. Robert Rogers, examining marl deposits near Richmond, wrote his brother William: "But I forgot to mention the most interesting organic remains we have yet met with—I mean the charming damsels we everywhere meet with. . . . We are at the moment staying with Mr. Brown, a very hospitable gentleman & he has two very pretty & interesting daughters."[26]

Each year, the annual reports showed some new discovery in Virginia geology or some development in William Rogers' thought. The 1836 report, published in 1837, still used the descriptive names of the 1835 reconnaissance, based on texture, color, mineral contents, and lithology. Most of the work centered on stratigraphy, not structure. This was an administrative report, heavy on economic geology and written from notes in travelogue style.[27] In 1837, Rogers had more assistants in the field. They were now working intensively in the mountains, not just getting acquainted with the order of the formations. The hardship of this sort of field work was described in the annual report, if not as graphically in the letters. Rogers was eloquent on the exasperation his geologists felt in struggling to use the official topographic map of Virginia as a base for their geologic map. Often rivers would be misplaced or streams missing. Mountain ridges and boundaries were sometimes four to twelve miles out of line. The 1837 report included a narrative on the discovery of a long fault line that upset the tidy sequence of the 1835 profile. The geologists kept running cross-sections rather than going along strike to trace formation boundaries. One party ran over twenty sections in their region. Rogers then had to piece all these sections together to map the state's geology. In 1837, the formations were numbered as well as described; this was the first publication that demonstrated the Rogers brothers' insistence that American geological investigations were too preliminary to warrant assigning formational names.[28]

In 1838 work in the Appalachians delayed investigation of the Piedmont and Tidewater for one season. The geologists refined their sections of previous years by doing altitude observations and eliminating vertical exaggeration. The report defined several structural terms—fault, anticline, syncline, and the like. Rogers recognized the concept of thrust faults, although he did not use that phrase. The scientists discovered an inversion of the normal order of formations in the North Mountain complex, but they did not say whether an overturned fold or a thrust fault accounted for it. During 1838 Rogers also confirmed that the setting properties of hydraulic cement were correlated with magnesia content.[29]

The 1839 report came closer to a modern geologic monograph than those of earlier years. It defined the rock terms in one section, for example, instead of spreading them throughout the report or leaving them undefined. For the first time, details of the state's Mesozoic formations were reported; their existence and depositional history had only been mentioned before. The more experienced assistants stopped running cross-sections and began to trace formation boundaries on the map. The geologists started work on the igneous rocks that clutter part of the Virginia landscape. And, insofar as this information could be conveyed verbally rather than with a map, they showed how geology controlled topographic relief. Finally, Rogers analyzed the coal basins in the western part of the state, in a part of the report that was a virtual microcosm of geosynclinal ideas. The 1840 report gave more details on all these phenomena and, in addition, announced Rogers' discovery of infusorial strata of Tertiary age in the state. This report also reprinted the glossary of terms from Lyell's *Principles*.[30]

In 1841 the legislature suspended the survey, effective in 1842, and the report for that year was quite slim and again administrative rather than geological in nature.[31] Rogers was sure he would finish the final report the next year, but the money to print it was not forthcoming, and even with adequate funding, it is doubtful that Rogers would ever have completed it. He was a geologist who always thought he needed just a little more field work before settling down to compose his magnum opus. This pattern of thought ran in the family; Henry Rogers took from 1842 until 1858 to issue his final report for Pennsylvania.

In defense of the Virginia legislature, it should be said that their suspension of the survey was due to economic difficulties centering on the recent national depression, not on political factors. Virginia carried a heavy interest payment for its transportation works and was loath to suspend payment if the state could economize by reducing or terminating nonessential programs such as the survey.[32] The failure of the survey to issue a final report meant that it in all likelihood would not receive, and indeed it has not received, the detailed attention awarded to New York or other states where the task was completed.

For Rogers, the intellectual culmination of his work came in an 1842 paper (published in 1843) on the structure of the Appalachians, which he wrote with his brother Henry. Patsy Gerstner traced the intellectual genesis of this document and attributed most of the theoretical parts on mountain origin to Henry Rogers.[33] However, the descriptive aspects, which outlasted the theory, certainly owe much to the Virginia work of William Rogers. In 1924 Merrill observed, "This work, so far as it related to the structure of the [Appalachian] chain, has been improved upon in recent times only by the discovery of enormous overthrust faults in the southwestern portion."[34]

As with the 1835 Virginia reconnaissance report, several geological cross-sections accompanied the paper on Appalachian structure. This time the sections were free of vertical exaggerations and used the Rogers' numbering system for formations in place of descriptive terms.[35] The illustrations for William Rogers' paper on thermal springs, published simultaneously in the same journal, included several local detailed sections (for example see Fig. 5B), also without vertical exaggeration.[36] One plate of sketches accompanying the text on Appalachian structure documented William Rogers' increased reliance on faulting to explain geologic configurations (see Fig. 5A), and two, one of which accompanies the text on thermal springs, emphasized the association of

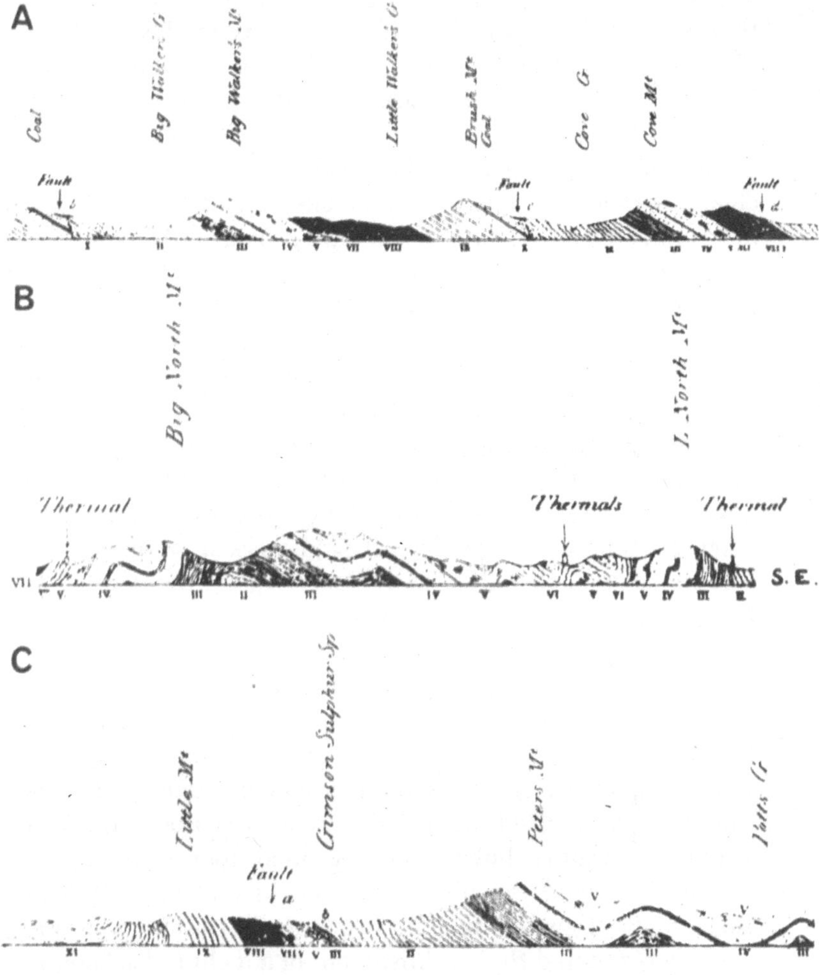

Figure 5.
Three sections, A and C taken from plate 21 of the Rogers brothers' publication on the physical structure of the Appalachian chain, and B taken from plate 16 of William B. Rogers' publication on the connection of thermal springs in Virginia with folds and faults. A illustrates faulting to explain geologic configuration. B shows the association of thermal springs with folds. C shows the association of a spring with a fault. Note lack of vertical exaggeration in these sections.

thermal springs with faults and folds (for examples see Figs. 5B and 5C), a phenomenon remarked upon frequently in the Virginia reports and summarized in the paper on thermal springs.

The Rogers brothers' 1842 (1843) paper on the Appalachians presented some of the details and ideas that might have appeared in a final report on the geology of Virginia, but it went beyond the conventions of state surveys of the time to place the state in a wider scientific perspective. However, Herbert Woodward, in 1961, selected 1838 as the turning point in Appalachian geology. He specifically referred to the section of William Rogers' Virginia geological survey report for that year titled "Geology of the Appalachian Region," as the "grandfather of all such [articles]" with which he was personally familiar. Woodward further observed that in 1925, when he first started geological field work in Virginia, "for various parts of that state and West Virginia there were no more recent detailed studies than Rogers' classic reports."[37]

The Virginia state legislature refused to renew funding for the continuation of the survey beyond April 1842. And, although Rogers retained the nominal title of state geologist, for practical purposes coordinated survey studies of the state ceased. In 1848 Rogers resigned as state geologist and at the same time submitted his resignation as professor at the University of Virginia, but he was prevailed upon to remain on the faculty, which he did until 1853. In that year, William Rogers left Virginia to take up residence in Boston, where he actively promoted the establishment of a technical school, an idea he and his brother Henry shared as early as 1846.[38] The dream became a reality in 1861, when the Massachusetts Institute of Technology was chartered by the state legislature, with William Barton Rogers as its first president. In 1863 Rogers joined forty-nine other "incorporators" in founding the National Academy of Sciences, of which he served as president from 1878 to 1882.[39] Rogers continued to be active in the scientific and educational communities until the day of his death, May 30, 1882, at the age of seventy-seven, while delivering the commencement address at MIT.[40]

It is certainly regrettable that William Barton Rogers' accomplishments with the Virginia geological survey during 1835 and the years immediately following have been obscured by the more visible results of the New York and other state surveys for which impressive final reports were issued. But William Rogers cannot be denied his rightful place in the annals of nineteenth-century North American geology and in the history of intellectual development of the period.

Notes

The authors thank John Salmon of the Virginia State Library and Karen Templeton of the Massachusetts Institute of Technology Archives for providing access to documents during visits to their respective institutions. They also express appreciation to Ellis Yochelson, U. S. Geological Survey, and Ray Brian, Librarian, California Academy of Sciences, for assistance in locating illustrations, and to Peter U. Rodda, Department of Geology, California Academy of Sciences, for reading and commenting upon the manuscript.

1. William Barton Rogers, "Report of the Geological Reconnoissance [*sic*] of the State of Virginia," Virginia State Legislature, House of Delegates Documents no. 24 (1836), pp. 1–52. Also printed as *Report of the Geological Reconnoissance of the State of Virginia* (Philadelphia: DeSilver, Thomas, and Company, 1836), p. 143 and reprinted in idem, *A Reprint of the Annual Reports and Other Papers on the Geology of the Virginias* (New York: Appleton, 1884).

2. Robert Rakes Shrock, *Geology at M.I.T. 1865–1965: A History of the First One Hundred Years of Geology at the Massachusetts Institute of Technology* (Cambridge: MIT Press, 1977), pp. 208–14, lists biographies of Rogers and histories of the Virginia survey.

3. On Henry Rogers, see Clark Elliott, *Biographical Dictionary of American Science: The Seventeenth through the Nineteenth Centuries* (Westport, Connecticut: Greenwood, 1979), pp. 218–19.

4. George Perkins Merrill, "Contributions to a History of American State Geological and Natural History Surveys," United States National

Museum Bulletin no. 109 (1920), pp. 45–46, 137–39, 149–52, 307–9, 363–68, 459–60, and 464–66.

5. On Browne and his group, see Anne Millbrooke, "The Geological Society of Pennsylvania, 1833–1836," *Pennsylvania Geology* 7 (1976):7–11 and *Pennsylvania Geology* 8 (1977):12–16.

6. Merrill, "State Surveys," pp. 507–8.

7. Ibid., p. 508.

8. Emma Savage Rogers, ed., *Life and Letters of William Barton Rogers* (Boston: Houghton Mifflin, 1896), 1:116–17.

9. Virginia State Legislature, House of Delegates Journal (1835), pp. 215–16.

10. Merrill, "State Surveys," pp. 508–9.

11. These papers are most accessible as reprinted in Rogers, *Geology of the Virginias*, pp. 1–20 and 659–74.

12. Julia Gardner, "Mollusca from the Miocene and Lower Pliocene of Virginia and North Carolina," United States Geological Survey Professional Paper no. 199 (1943), p. 3, places Rogers' work on the Tertiary in the context of earlier and later investigators. Dr. Lyle Campbell of the University of South Carolina at Spartanburg supplied this reference.

13. Patsy Gerstner, "Henry Darwin Rogers and William Barton Rogers on the Nomenclature of the American Paleozoic Rocks," in *Two Hundred Years of American Geology: Proceedings of the New Hampshire Bicentennial Conference on the History of Geology*, ed. Cecil Schneer (Hanover, New Hampshire: University Press of New England, 1979), pp. 175–86.

14. Benjamin Silliman, "Remarks on Some of the Gold Mines and on Parts of the Gold Region of Virginia. Founded on Personal Observations Made in the Months of August and September, 1836," *American Journal of Science*, 1st ser. 32 (1837):98–130.

15. John T. Lonsdale, "Geology of the Gold-Pyrite Belt of the Northeastern Piedmont Virginia," Virginia Geological Survey Bulletin no. 30 (1927), p. 11.

16. Rogers, "Reconnoissance," pp. 51–52; idem, *Geology of the Virginias*, pp. 120–22.

17. Merrill, "State Surveys," pp. 509–10.

18. Rogers, *Life and Letters*, p. 130.

19. See especially the letters to William Rogers from Charles B. Hayden (1836–38), Israel Slade (1838–40), William Aikin (1838), James B. Rogers (1838–39), and Caleb Briggs (1839–41) in the Virginia Geological Survey Papers (Accession 24815), Archives Division, Virginia State Library, Richmond, Virginia. There are also revealing letters on survey fieldwork in the William Barton Rogers Papers (MC-1), Institute Archives and Special Col-

lections, Massachusetts Institute of Technology Library, Cambridge, Massachusetts, notably those from Robert E. Rogers and James B. Rogers in folders 13–18 (1836–41 inclusive).

20. Boyd's letters for 1838–40 are in the Survey Papers, Virginia State Library.

21. Caleb Briggs to William Barton Rogers, October 8, 1840, Survey Papers, Virginia State Library.

22. William Barton Rogers to Henry Darwin Rogers, December 22, 1840, Survey Papers, Virginia State Library.

23. Hayden to William Barton Rogers, May 19, 1838, Survey Papers, Virginia State Library.

24. Hayden to Rogers, August 25, 1838, Survey Papers, Virginia State Library.

25. Mark Aldrich, "Earnings of American Civil Engineers, 1820–1859," *Journal of Economic History* 31 (1971):407–19.

26. Robert E. Rogers to William Rogers, May 14, 1836, Survey Papers, Virginia State Library.

27. The annual reports were issued in three formats. They first appeared in the Virginia State Legislature, House of Delegates Documents series; then as William Barton Rogers, *Report of the Progress of the Geological Survey of the State of Virginia for the Year 1836* and subsequent years through 1841 (Richmond: Samuel Shepard, 1837–42); and finally reprinted in *Geology of the Virginias*. The last version is available in most libraries and will be the form cited hereafter.

28. Rogers, *Geology of the Virginias*, pp. 147–186; Gerstner, "Nomenclature."

29. Rogers, *Geology of the Virginias*, pp. 189–243.

30. Ibid., pp. 245–535.

31. Ibid., pp. 537–46.

32. See especially the correspondence between J. Brown of the Board of Public Works and W. B. Rogers in Survey Papers, Virginia State Library; the correspondence relating to the survey in Box 51 of the Board of Public Works Records, Archives Division, Virginia State Library; and the letters for 1842 in folder 19 of the Rogers Papers (MC-1) at MIT.

33. Patsy Gerstner, "A Dynamic Theory of Mountain Building: Henry Darwin Rogers, 1842," *Isis* 66 (1975):26–37.

34. George P. Merrill, *The First Hundred Years of American Geology* (New Haven: Yale University Press, 1924), p. 219.

35. "On the Physical Structure of the Appalachian Chain as Exemplifying the Laws Which Have Regulated the Elevation of Great Mountain Chains Generally," *American Association of Geologists and Naturalists*

Transactions, 1840–1842 (1843), pp. 474–531; reprinted in Rogers, *Geology of the Virginias*, pp. 599–642.

36. "On the Connection of Thermal Springs in Virginia with Anticlinal Axes and Faults," ibid., pp. 323–47; reprinted in *Geology of the Virginias*, pp. 575–99.

37. Herbert Woodward, "Reappraisal of Appalachian Geology," Bulletin of the American Association of Petroleum Geologists Bulletin, 45 (1961): 1626.

38. Merrill, "First Hundred Years," p. 186.

39. Mary Rabbitt, *Minerals, Lands, and Geology for the Common Defence and General Welfare*, vol. 1, *Before 1879* (Washington, D.C.: U.S. Government Printing Office, 1979), p. 144.

40. Joseph K. Roberts, "William Barton Rogers and His Contribution to the Geology of Virginia," Virginia Geological Survey Bulletin no. 46 (1936): 25–28.

Southern Influences on the Career of Joseph Nicollet

Martha Coleman Bray

Joseph Nicollet (1786–1843), French astronomer and cartographer, is known today especially for his map, the "Hydrographical Basin of the Upper Mississippi River," published by the U.S. Senate in 1843 and the U.S. House of Representatives in 1845. Nicollet arrived in the United States early in 1832, unheralded but not unknown. A discoverer of comets, a surveyor of distinction, and a contributor to the fields of geodesy and magnetism, he had worked in the Paris Observatory and in the Bureau of Longitudes. His purpose in coming to America was precisely stated: to make "a scientific tour, with the view of contributing to the progressive increase of knowledge in the physical geography of North America."[1] It soon became clear that a geographical study of the Mississippi River was "the reason and interest of all his work."[2]

Nicollet left Paris under a cloud of financial and political difficulties after the revolution of 1830, but he had known for some time that his observational skills were needed nowhere so much as on the continent of North America. Nicollet was unusually well endowed with both geographical and astronomical skills, but his only formal exposure to geological thought seems to have been through the lectures of Lieutenant Rozet of the Royal Corps of Geographical Engineers, whose *Elementary Course in Geognosy* was published in 1830.[3]

Although Nicollet's professional experience and his advanced learning were acquired in Paris, he retained many of the attitudes and values of a less urbane, less urban culture. He was very much the product of the mountain province of Savoy, where he was reared and where he received his early education. In eighteenth- and early nineteenth-century Savoy, the best seventeenth-century Jesuit traditions prevailed. It may be said that Nicollet's scientific "pilgrimage," as his friends liked to say, spanned not only an ocean and two continents, but also two centuries of cultural and intellectual growth.

The geological contributions that were to raise Nicollet to the front rank among American geologists were made during two expeditions in 1838 and 1839, under the Corps of Topographical Engineers. With John Charles Frémont as his assistant, Nicollet eventually mapped the triangle of land between the Mississippi and Missouri rivers. His travels in the South, from 1832 through 1835, were consistently and progressively a preparation for these later contributions.

Nicollet's meticulous notebooks record all of his scientific observations from Baltimore to Savannah; across the mountains of Georgia to Nashville, Tennessee; northward to the Ohio River; then down the Ohio, eventually reaching the Gulf of Mexico. These notebooks also tell us, both directly and indirectly, of a mind open to new impressions and of a cultural climate particularly able to enrich an inquiring mind.

As a scientist and as a thinker, Nicollet's classical training in astronomy and mathematics was always tempered by a Humboldtian concept of physical geography. Though he was known only slightly to Humboldt during the baron's triumphant years in Paris, the young man from Savoy can be said to be one of the geographer's truest disciples. What Hans Baumgartel has said of Alexander von Humboldt could also be said of Nicollet: "His nature was to connect ideas, to perceive relationships," and to believe passionately in "the material unity of the world and the connection of all of the laws of nature."[4] Geology, particularly as it was evolving in the early nineteenth-century United States, was a natural stimulus to such a mind. The line between physical geography in its broadest aspects and the geological observations made in America up to 1832 was still barely distinguishable. As state geological surveys got under way, the first of them in the South, the need for mapping grew more imperative.[5] Nicollet eventually helped to meet this need.

As Nicollet traveled throughout the South, he moved from home to home within one well-defined social circle. The society within which Nicollet lived was a French society. The people, if not entirely French in origin, were French by influence and sympathy. It is remarkable to find an intellectual Frenchman straight from Paris learning from new friends in the United States what had been going on in scientific circles abroad. Aside from this Frenchness, Nicollet's hosts welcomed him as a friend in need of shelter and as a "man of theory" who was willing to devote himself to matters of utmost practical importance. It was wonderfully liberating for a man who had been the victim in Paris of a centralized, rigidly politicized, and nationally powerful scientific elite to discover everywhere he went that the observational sciences, controlled by fact rather than speculation, were of supreme interest. Throughout his notes and letters and in his published report there runs the constant theme of gratitude to the "voluntary assistants" he found "among the enlightened men who are spread in universities, in colleges and in society," and the "liberal and disinterested intercourse which characterizes American savants."[6] Unlike many European observers, he was always helpful, optimistic, and encouraging.

From Washington, where he made useful acquaintances but obtained no solid support from the government for his proposals, Nicollet went to Baltimore, where, Roman Catholic that he was, he found a home among the Sulpician fathers of St. Mary's College. With these democratic and scientifically intelligent clergymen he shared in the common interests of a community where canals, railroads, and other internal improvements loomed large. In Maryland Nicollet began a series of precisely located, fixed stations, where he compiled geographical and astronomical data based on careful instrumental observations. Wherever he traveled in North America, Nicollet continued to establish these fixed stations. Most of his observations were barometric measurements of the elevation of the land.

Like the rest of Nicollet's southern hosts, the French population of Baltimore was largely composed of families who had fled from the slave uprising in San Domingue, or what is now Haiti and part of the Dominican Republic. They were professional people who settled throughout the southern states to contribute their skills and knowledge as citizens of the Republic. It is a temptation to expand upon this unique society of Baltimore, but here we must concentrate on Jules Timoleon Ducatel, Nicollet's lifelong friend.

Ducatel was the son of a French pharmacist. He had been educated at St. Mary's College, Maryland, and had studied in France, returning to Baltimore in 1824 with an enthusiasm for chemistry and geology. Ducatel was not the only citizen of Maryland who was deeply impressed with French scholarship in geology. For what it is worth as a social comment, Judge William Wirt, who upheld the case of the Cherokee Indians before the Supreme Court, wrote in that same year in a letter: "Have you ever read Cuvier's theory of the earth? Here is a book which has been before the American world for eight or ten years, full of the most interesting and instructive facts and speculations, and yet within a fortnight past, I have known nothing of it, save only the name. This same geology, with all its discoveries of the past history of our planet, is a prodigious science, and nearly as honourable to the sagacity of man, in my opinion, as astronomy."[7] Judge Wirt's son-in-law, Lewis Goldsborough, became one of Nicollet's many young disciples. He was Nicollet's helper, host, and confidant in Florida.

In 1832, Ducatel was appointed professor at the respected Medical College of the University of Maryland, but his enthusiasm for geology was so great that his friends of the Delphian Literary Club gave him the nickname of *Basaltes craniosorsum*. A few months after Nicollet left Baltimore, Ducatel and a twenty-year-old surveyor, John Henry Alexander, who was later to prepare the mathematical portion of one of Nicollet's reports for publication, were appointed to make a geological survey and a topographic map of Maryland. This was one of the first state surveys to emphasize a topographic map, and the youthful Alexander was the topographer.

As Nicollet was to learn, travel in the South was difficult. Roads were poor, bridges were scarce, and the generally accepted distances between towns were often rough approximations. Thus it is not surprising that the Maryland survey, with its focus on topography, was directed by the agency responsible for public transportation. From the general state of geographical knowledge, it is also clear that Maryland's first topographer faced a great challenge. Alexander was quite successful and his letters reveal a close relationship between his methods and those of Nicollet.

In 1832 Nicollet left Baltimore to begin his exploration of the continental interior. Cholera on the river steamboats sent him to Charleston rather than to the Mississippi, as he had originally planned. Thus, Nicollet began a sojourn in the southern states that lasted until 1835 (Fig. 1). The years spent in the South strengthened Nicollet's original purpose rather than diminishing it. He remained intent on mapping the Mississippi Valley, and the South became a testing ground for Nicollet's skills as a field worker.

In making scientific observations like those required to calculate longitude, Nicollet badly needed assistants. The

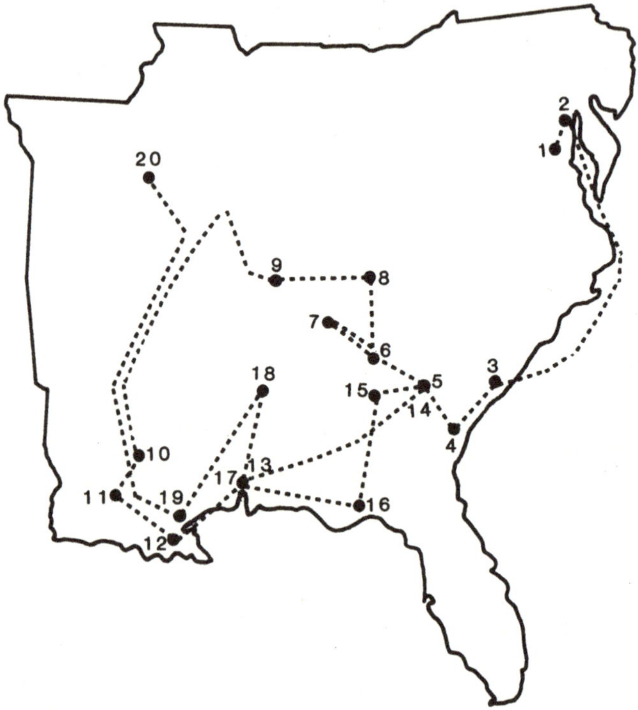

Figure 1.

NICOLLET'S SOUTHERN TRAVELS, 1831–35. (1) Washington, D.C., 1831; (2) Baltimore, Maryland, 1831–32; (3) Charleston, South Carolina, 1832; (4) Savannah, Georgia, 1832; (5) Augusta, Georgia, 1833; (6) Athens, Georgia, 1833; (7) Lookout Mountain, Georgia-Tennessee, 1833; (8) Great Smoky Mountains, Tennessee, 1833; (9) Nashville, Tennessee, 1833; (10) Natchez, Mississippi, 1834; (11) Washita River, Arkansas, 1834; (12) New Orleans, Louisiana, 1834; (13) Mobile, Alabama, 1834; (14) Augusta, Georgia, 1834; (15) Milledgeville, Georgia, 1834; (16) Tallahassee, Florida, 1834; (17) Mobile, Alabama, 1835; (18) Tuscaloosa, Alabama, 1835; (19) New Orleans, Louisiana, 1835; (20) St. Louis, Missouri, 1835. Nicollet's very detailed notes generally give dates of arrival and departure. The notes also identify people Nicollet met and provide exact locations for fixed stations where Nicollet determined the elevation of the land, made astronomical observations, calculated longitude, collected rocks, or made other observations that he considered to be significant.

huge valley of the Mississippi and its tributaries extended westward from the Alleghenies. Nicollet's area of interest included the watershed of the Tennessee River and the Gulf Coast as far east as Florida.[8] To deal with this immense area Nicollet needed many helpers. Initially, assistance with southern observations came from the faculty of a medical college in Augusta, Georgia, where three remarkable men, Milton Anthony, Lewis D. Ford, and especially Louis Alexander Dugas, joined his growing force of observers. Dugas was of a San Domingan family.[9] He had received help with his education from a French émigré, Dr. Charles L. de Beauregard. Dugas studied medicine in Baltimore before sailing for Paris in 1828. In Paris he had a full course of lectures, including Beaumont on geology. While Dugas is primarily remembered as a surgeon, in 1832 he was singularly well qualified to assist Nicollet in general scientific observations.

After working in the Augusta area, Nicollet moved on to Athens, Georgia, drawn as a topographer toward the mountains. In Athens he had established contact with several faculty members at Franklin College. With Dr. James Jackson, professor of French and geology, and Henry Hull, professor of mathematics, he proceeded into the high country around Clarksville, Georgia, surveying Currahee Mountain. Later, with a new entourage that included a French teacher from Savannah, he proceeded across a wild stretch of country to Lookout Mountain, on the Tennessee border.

He described the view from Lookout Mountain and made surmises as to the unity of the whole Appalachian chain.[10] Because of his observations on the Georgia-Tennessee border, Nicollet was an enthusiastic spectator at the 1842 meeting of the American Association of Geologists, where the Rogers brothers, two renowned geologists he had met in Baltimore, presented a paper on "the physical structure of a mountain chain 1,300 miles in total length . . . and 100 miles in its greatest breadth." He fully concurred in the statement that "a grander geological theme could hardly be imagined," and spoke enthusiastically of it to their mutual friends.[11]

In Georgia, Nicollet's friends urged him to stay. They were desperate to save their state from what Nicollet recognized as the "premature discovery of gold." Horrifying bills were being enacted by the legislature to aid the greedy, as one glance at the *Southern Banner* of 1833 will show. Intelligent Georgians questioned state support for speculative mineral industries. They preferred to see public money spent for better roads and similar internal improvements. Those who opposed the gold promoters were politically active, and Nicollet joined them, writing a letter to Congressman Richard Henry Wilde in which he echoed their antigold sentiments.[12]

Among Nicollet's papers is a map entitled "Geological Map of the Mining Districts of the State of Georgia and the Western Part of North Carolina and Tennessee by Jacob Peck, entered according to Act of Congress." On the map, various geological formations are vaguely defined, and the word GOLD glitters as a lure to the inaccessible county. It was this sort of cartography that prompted a wandering French scientist to become involved in American political affairs.

Despite the urging of friends, Nicollet soon moved on toward the mountains of eastern Tennessee. His fascination with high places was more than a love of scenery. To the twentieth-century scientist aided by aerial photographs and fine maps, it is difficult to appreciate the world of the nineteenth-century foot-traveler. Nicollet traversed a landscape almost devoid of precisely located topographic and cultural features. There were few exactly measured elevations, none on the frontier. The topography he studied had no precisely measured distances and no astronomically determined locations. For Nicollet, the view from the mountain top put his work in natural perspective.

Nicollet's field notes describe him standing on one of the

highest summits in the Smoky Mountains, with swollen feet, shivering in the cold. His precious barometer hung on a tree and he stood looking down from the mountain top, eating raw chestnuts and struggling to clarify the geographic relationships of the country he was exploring. In eastern Tennessee and adjacent areas, the view from the mountain top was essential. As he once wrote in a letter, it was not enough "to wander through the lower valleys, picking up flowers, plants and stones."[13]

From eastern Tennessee, Nicollet made his way to Nashville, arriving in a state of physical and mental exhaustion. While he had not originally planned a lengthy stay, Nicollet spent about six weeks in the company of one of Tennessee's best-known intellectual leaders, the French-speaking Gerard Troost. This cheerful and energetic Dutchman's background and his work with the Tennessee geological survey are well known to students of the history of geology. A few quotations from five long, chatty and affectionate letters which Troost wrote to Nicollet between 1835 and 1843 will show that their meeting was of benefit not just to the Frenchman.[14] Troost was lonely for intellectual companionship and had enormous enthusiasm for American geology. Their interaction changed Nicollet's awareness of topography into an earnest, working observation of the rocks along his route. If Troost was Nicollet's tutor, Nicollet was his companion, and they formed a solid friendship.

Nicollet's letters to Troost are apparently lost, but many of Troost's letters survive. In August of 1834 Troost wrote describing his examination of the Cumberland Mountains from Huntsville, Alabama, to the Cumberland River. He writes as to a well-trained geologist: "As you already know the geological environs of Nashville . . . I need not describe it." He went on to say that ten to twelve miles north of Huntsville "I saw for the first time the true place of *our* oolitic limestone. It is under the coal formation." And in the same letter, "Well, my conjectures are proving to be true in regard to the junction of the graywacke formation and the mountain limestone." And, "My excursion of the autumn ought to be in the western district. I only wish I could have the pleasure of making this excursion in your company." On December 28, 1835, he wrote to Nicollet in St. Louis: "I hope you are going to the neighborhood of Michel where there are near the sources of the St. Francis River, the beautiful porphyres and lignites and the trap rock . . . subordinated to the syenite." And a passage that has been quoted before, but must be quoted again: "I hope that you will have several new fossils to show me and that I may describe them. And the minerals to examine. When you give me the pleasure of passing some time here, bring everything you think might be new as you will lodge in our house where you will have a room 25 feet square so that we can pass the days and part of the nights together. And we will be able to fossilize, mineralogize, crystallographise and rocify."

Among the technical terms in these letters is one nongeological word that defies research thus far. Glad to hear that Nicollet has survived a summer in New Orleans and encouraging him to visit Tennessee, Troost wrote, "We will sing *latificamus* and on the first occasion I will drink a *tongebord* to your health." What is *tongebord*?A new drink? Or a form of toasting?

From Nashville, Nicollet made his way by steamboat up the Cumberland, down the Ohio, and in below-zero weather turned south to Natchez. There he stayed again with a French Dominican family, that of Henry Chatard Laplace, who had friends in Baltimore. Mired in mud and rain, Nicollet studied the work of William Dunbar, whose son lived on a neighboring plantation. Dunbar, a friend of Thomas Jefferson, had carefully observed the tides of the Gulf of Mexico.[15] Later, Nicollet was himself able to observe the fluctuations of the level of the river as it entered the Gulf, which Dunbar had observed to occur once every twenty-four hours and approximately at the same time every morning. Nicollet also studied the work of Andrew Ellicott, and cherished a statement he had found in a letter from Ellicott to Dunbar: "Truth will triumph eventually in a free government."[16] The rest of the letter urged greater government support for science, scorning false economies. Nicollet, like Ellicott, was interested in ground-water temperature and in the shifting meanders of the Mississippi. His observations of the shifting passes between the fingers of the Mississippi Delta, based upon historical records, reveal a mind well prepared for historical geology.

The effects Troost's teaching had on Nicollet became apparent on the thirty-first of March 1834, when he stepped from his steamboat and climbed a fifty-foot outcrop of green clay on the right bank of the Washita River, a spot accurately located in his notebook. There, Nicollet carefully collected a Tertiary fossil, which, with others from other localities, was eventually sent to the noted Timothy Conrad for identification and description.[17]

All of Nicollet's peregrinations in the South do not concern us here, but the French influence, if not the geological, traces a cultural pattern that has not received the attention it should. A tantalizing and surprising visit to Tuscaloosa mentions his observations on the location of a coal mine there, in which he unfortunately lost his chronometer. Again, his guide was French, this time by birth, one Sauvier F. Bonfils, a professor of modern languages at the University of Alabama.[18]

Nicollect reached St. Louis in 1835. His meeting there with George Featherstonhaugh, who had just returned from his expedition across what was then Wisconsin Territory, was the second watershed in his American geological education.

Urged by Troost and stimulated by Featherstonhaugh's discussion of the European "Old Red Sandstone," Nicollet explored the Ozark region of Missouri. Entries in his notebook show he was well acquainted with Featherstonhaugh's expedition of 1834 to the country between the Missouri and Red rivers. He also took time to explore the Territory of Arkansas, following the Arkansas River to Little Rock and then going on to the Ouachita Mountains. On this trip he was accompanied by a new friend, Rush Nutt, of Rodney, Mississippi. Nutt's chief claim to fame is an important improvement in the cotton gin. Few people realize he also had geological interests.

While Nicollet's contact with Featherstonhaugh was generally stimulating, it was as a cartographer that Featherstonhaugh most particularly stirred the Frenchman's interest. Contact with Featherstonhaugh focused Nicollet's attention on the Côteau des Prairies, that extraordinary rise of land which is the watershed between the Missouri and the Mississippi. This was the area that eventually became the heart of Nicollet's hydrographical map. The four southern years were important preparation for this major mapping project and for other contributions that Nicollet was later able to make to American geology. Nicollet had gradually gained what might be called a continental concept—a general view that was essential to his mind, a mind that his friends so often referred to as philosophical. Through French-speaking connections he had caught up with knowledge denied to him in Paris, and he had acquired an admiration and familiarity with American institutions of learning as well as friendships with American intellectuals. His southern travels also afford a brief glimpse into the rich and varied history of local associations and institutions that make the early nineteenth century in the United States such an interesting chapter of intellectual and cultural history.

Notes

1. *Report Intended to Illustrate a Map of the Hydrographic Basin of the Upper Mississippi River,* made by J. N. Nicollet, while employed under the Bureau of the Corps of Topographical Engineers, 26th Congress, 2d Session, Senate Document 237, serial 380 (Washington, D.C., 1843), p. 3.

2. Nicollet to Samuel Eccleston, March 10, 1834, Eccleston Papers (Archives Archdioces of Baltimore).

3. *Cours Elementaire de Geognosie,* by M. Rozet, Lieutenant of the Royal Corps of Geographical Engineers, Member of the Society of Natural History of Paris, etc. (Paris, 1830).

4. Hans Baumgartel, "Alexander von Humboldt," in *Toward a History of Geology,* ed. Cecil J. Schneer (Cambridge, Mass.: MIT. Press, 1969), pp. 20–21.

5. George Merrill, "Contributions to a History of American State Geological and Natural History Surveys," Smithsonian Institution, U.S. National Museum Bulletin no. 109 (Washington, 1920), p. 139; Michele Aldrich, "American State Geological Surveys, 1820–1845," in *Two Hundred Years of Geology in America,* ed. Cecil J. Schneer (Hanover, N.H., 1979), pp. 133–43.

6. Nicollet, *Report,* p. 4.

7. J. P. Kennedy, *Memoirs of William Wirt,* 2 vols. (Philadelphia: Lea and Blanchard, 1849), 2:180.

8. Timothy Flint, *A Condensed Geography and History of the Western States or the Mississippi Valley,* 2 vols. (Cincinnati: E. H. Flint, 1828), 1:18–19.

9. W. H. Goodrich, *The History of the Medical Department of the University of Georgia* (Augusta, n.d.), pp. 228–31.

10. Nicollet Papers, Library of Congress, Astronomical Notebook, April 15, 1833. All references by Nicollet to his route and to incidents and friends are from this source.

11. Emma Savage Rogers, *Life and Letters of William B. Rogers,* 2 vols. (Boston: Houghton Mifflin, 1896), 1: 120, 121, 124.

12. Nicollet to Richard Henry Wilde, August 28, 1833, Greer Papers, Historical Society of Pennsylvania.

13. Nicollet to Eccleston, September 27, 1833, Eccleston Papers.

14. Troost to Nicollet, August 12, 1834; December 28, 1835; November 16, 1840; Nicollet Papers, Library of Congress. Troost to Nicollet, October 1835, Miscellaneous Papers, New York Public Library. Troost to Nicollet, Sept. 8, 1841, Nicollet Papers, Minnesota Historical Society.

15. William Dunbar, "Description of the River Mississippi and its

Delta," *Transactions of the American Philosophical Society* 6, pt. 1 (1809): 165–87.

16. Andrew Ellicott to Dunbar, January 13, 1801, copy in hand of Nicollet, Nicollet Papers, Library of Congress.

17. Martha C. Bray, "Joseph Nicolas Nicollet, Geologist," in *Proceedings of the American Philosophical Society* 114, no. 1 (1970):41.

18. *American Biographical Notes*, arr. Franklin B. Hough (Albany: Jack Munsell, 1875), p. 35.

Antebellum Geological Surveys in Kentucky and Their Contribution to the Shaler Survey of the 1870s

Ivan L. Zabilka

The legislature of the Commonwealth of Kentucky authorized statewide geological surveys relatively late in the antebellum period and provided little financial support for them. Fiscal necessity rather than vision dictated the size, scope, and timing of efforts to survey the state. The thrust of geological work prior to the Civil War was practical rather than scientific. While records are scant, they allow at least a limited reconstruction of the historical setting of the early Kentucky surveys.

The earliest literature on exploration and travel in the state provides occasional notations concerning geological features. Topography, mineralogy, and paleontology were sometimes mentioned in the notes of the naturalists of the day. Thomas Jefferson, for example, wrote of large mammalian bones found at a salt lick in the northern part of the state near Cincinnati. One of the earliest serious commentaries upon Kentucky landscape features was by the famed explorer John Filson. In 1784 he published a book about Kentucky in which he described the spectacular 300-foot cliffs, the Palisades, along the Kentucky River, bordering the present Fayette and Jessamine counties.[1] Filson further described the salt springs of Drennon and Big Bone, and the presence of lead, iron ore, and crude oil.

Jefferson and Filson were not the only serious writers and travelers who commented upon Kentucky. The British naturalist Gilbert Imlay in 1792 provided "a number of very clear and well written passages on some aspects of regional topography in Kentucky, particularly a paragraph or two on the Barrens of southwestern Kentucky and the sinks and caves that mark this remarkable area."[2] In addition, other works providing short but reasonably accurate statements concerning geological features include those of Thomas Cooper, published at Dublin in 1794; James Meace, at Philadelphia in 1807; William Maclure, at Philadelphia in 1809; Jeremiah Rensselaer, at New York in 1823; and John J. Audubon, at Edinburgh in 1831.[3]

The City of Lexington, and more specifically Transylvania University, served as the intellectual center of Kentucky during the early nineteenth century. As news of geological activity in Massachusetts and New York, and then more immediately in Ohio and Indiana, reached Lexington, pressures and petitions upon the legislators grew. There was a grassroots demand for some effort to search out the riches of Kentucky. The result was the initial formal step toward a thorough geological account in the form of authorization for a reconnaissance. The legislative action has been described with these words: "On December 12, 1837, Senator Wingate and Representative J. T. Morehead presented memorials to their respective houses of the Kentucky Legislature requesting passage of a law authorizing a Geological Survey of the State. These memorials resulted in resolutions approved by Governor [James] Clark, which led to the first official . . . investigations in Kentucky."[4] The reconaissance was conducted in the summer of 1838 by William Williams Mather, who was then the director of the Ohio survey. He issued a forty-page report and encouraged the organization of a formal survey, advice that was not accepted by the legislators.[5]

Mather's report helped attract an increasing stream of

investigators who evaluated specific geologic features in which they were interested. Over the years James Hall, Timothy A. Conrad, Byrem Lawrence, Jules Marcou, David Dale Owen, Joseph G. Norwood, David Christy, Benjamin Silliman, and Sir Charles Lyell were among the naturalists and geologists who produced chapters, pamphlets, or other published comments about Kentucky geology. In addition the Commonwealth occasionally was mentioned in comparative accounts where information gathered in other states was the primary focus of study. This kind of treatment emphasized the need for a coordinated survey and created a sense that Kentucky was falling behind the other states. However, for fifteen years agitation did not generate sufficient interest or political support to warrant the investment of state dollars in a geological survey.

Following increased pressure from merchants, on March 4, 1854, a bill passed the legislature providing for a "geological and mineralogical survey."[6] The wording of this bill delineated practical and essentially nonscientific objectives for the newly created agency. Scientific aspects of survey work were further inhibited by the scale of the work and the level of funding. All of Kentucky was to be surveyed for a few thousand dollars. The legislature hoped the survey would lead to great discoveries of exploitable mineral wealth.

Kentucky was fortunate to secure the services on April 26, 1854, of the able David Dale Owen to direct the new survey. Owen, a noted Indiana geologist, had experience with several state geological surveys. He also had exceptional organizational ability. He needed both experience and ability, for he was granted just $10,000 to complete the huge task of surveying an entire state.[7]

Owen appointed Robert Peter of Transylvania University as chemist, and Sidney S. Lyon as topographer. From 1854 to 1859 this team, with additional help from Leo Lesquereux and Joseph Lesley during the last four years, did the field work upon which were based four published volumes, totaling over 2,000 pages. Owen provided direction, broad introductory chapters, and some field assistance during the first three years. Additional grants totaling $45,000 were made over the years until 1858. After October 1857 Owen worked primarily upon the survey of Arkansas, but did continue to provide some supervision in Kentucky. With Owen's death on November 16, 1860, publication of the final volume fell upon the shoulders of Robert Peter. Peter managed to get the volume through the press in the first year of the Civil War, but all other activity ceased.[8]

From the very beginning, Owen indicated that heavy attention would be devoted to coal and iron ore deposits.[9] The first step was to establish roughly the boundaries of these deposits, so the surveyors immediately turned to those portions of the state known to have mineral deposits. During the first field sessions of 1854 and 1855 portions of sixty-three counties were traversed in search of locations needing more attention. Robert Peter analyzed the composition of iron ores from various sites and even evaluated slags from functioning furnaces in an effort to aid that industry. Owen suggested means of more efficient operation in several cases. While the assistance given iron producers was significant, it was not sufficient to prevent the financial failure of small Kentucky furnaces, thus gradually ending local iron production. Improved technology and better ores from the Great Lakes forced an economic change.

Owen did not, however, interpret his mandate from the legislature narrowly. In the case of strong "magnesia" content in spring waters, Owen ventured into comments upon the health of livestock and humans, especially with regard to cholera and what he called milk sickness. A variety of rocks were subjected to chemical tests, with the intent of providing information about what could potentially prove useful in building. Dr. Peter also provided numerous analyses of soil samples collected in those parts of the state where the survey parties traveled. Finally, Owen made special note of bone deposits and of the many caves in the west central portion of the state, where the famous Mammoth Cave is located. While his primary focus was necessarily economic, Owen sought to draw attention to unique geologic features of the state that could be of interest to scientific observers.

The least effective and most disappointing portion of the report related to the topographical work. Only seventeen pages were needed to summarize what Sidney Lyon's survey team had accomplished. This summary consisted primarily of descriptions of points of prominence in the two counties where they had worked.

The second volume of reports covered the field work accomplished in the summers of 1856 and 1857. Owen's portion of the report was much briefer and related primarily to continued explorations of the coal and iron ore beds. The significant difference in this second report was the expansion of Peter's chemical reports, to nearly 170 of the 400 pages. Great attention was devoted to agriculture, with soil samples gathered along with control samples that demonstrated what nutrients were being removed from the soil by farming. The intent was to present the information in a fashion that would encourage the improvement of agricultural methods. Sidney Lyon's report was also much longer. He described the topographical work in Greenup County in the east and Hopkins County in the west, as well as prominent features in Crittenden, Caldwell, and Carter counties.

The third volume of reports was also based upon field work done during 1856 and 1857. This volume was late in production because of the accidental death of one of the field leaders. The major accomplishment presented in this volume was the more definitive location of the boundaries of the eastern coal field in Rowan, Bath, Powell, Estill, and Madison counties. Robert Peter continued his analyses of soils, wellwaters, and mineral springs. A basic change in the third volume was an increased attention to stratigraphy. The focus of stratigraphic study was an area in central Kentucky where two rivers, the Kentucky and Dick's (Dix), develop deep gorges and provide access to bold outcrops.

In the midst of this report, Owen provided a glowing discussion of the intended High Bridge of the Lexington and Danville Railroad. The bridge was to be a 1,250-foot suspension span some 400 feet above the bed of the Kentucky River. While the suspension never materialized, a steel bridge was built. This spectacular structure is still in daily use by the Southern Railway. It is the second highest railroad bridge in the eastern United States.

The topographic effort continued to provide the biggest headache to Owen and his teams. Owen wanted accurate levels in the coal fields, but lacked the means and funding to accomplish this. Levels were most frequently made with the barometer, an obviously inexact means. Owen also sought to establish an east-west baseline from which an extensive network of triangulation could be established. Brush and timber removal along this line greatly slowed the work of the field parties. Owen was forced to concede that triangulation in much of the state could not go forward until more of the heavy forests were removed. An added feature of the third volume of reports was the first extensive paleontological effort of the survey. Leo Lesquereux prepared a very lengthy, systematic description of fossils discovered by field parties. Most were plants and mollusks.

The final volume of reports appeared in 1861 and covered the field work of 1858–59. The base line was compiled from the Virginia border to the mouth of Highland Creek on the Ohio River, a distance of 306 miles.[10] The boundaries of the coal fields were essentially completed and initial maps were published, although these were not included with the reports. In the final report Owen indicated the counties that the survey had most neglected, and for the first time he complained about the lack of funds to complete field work and publish observations.

Some rumblings from the eastern academic establishment concerning Dr. Peter's chemical analyses caused Owen to present an extended defense. Questions seemed to relate primarily to the reliability of measures of trace elements and how so many analyses could be accomplished. Owen described in detail the analytical procedures so that others could judge for themselves the validity of the work done. Owen also took issue with European scientists who claimed there was a level below which further soil depletion would not take place under cropping. Peter's studies showed that soil exhaustion was taking place and that tillage alone could not replenish the soil, as was being claimed. Owen and Peter recommended fallow periods and the application of manure. Owen's final claim was that Kentucky's soils and ores had been more thoroughly analyzed than any comparable area of land in the world. This may have been true, if not for the world then at least for the United States.[11]

Leo Lesquereux continued his report, concentrating upon the fossil flora and the stratigraphic distribution of coals. Joseph Lesley reported upon the western edge of the eastern coal field. His report contains the first use of railroad surveys and rail bed cuts to augment the work of the state survey. As the extension of the railroads was promoted during the Civil War, this method greatly expanded in the second survey under Shaler. The railroads provided exposed strata and more accurate levels than the state survey had been able to achieve. Sidney Lyon provided a final topographical report, and the volume also contained a seven-page obituary of David Dale Owen.

In general, the first Kentucky survey touched upon the topography, stratigraphy, chemical analysis, mineralogy, and paleontology of the state. The value of the survey was uneven. Most disappointing was the topographical work. In large part this work had to be redone by the Shaler survey. On the other hand, since no adequate map of Kentucky existed in the 1850s, the survey provided a special service in producing the first reasonably accurate map of the major features with respect to distance and direction from each other. Further, the stratigraphic sections of the eastern and western coal measures rendered by Owen have never had to be revised significantly. These achievements aided the economic growth of the state, but there seems to be no way to measure the impact objectively and accurately in terms of money.

With the return of some modicum of normality after the war, interest in state geological efforts was renewed. Reconstruction caused economic adjustment. As steam replaced sail at sea, the demand for Kentucky hemp declined. Light industry was growing, but farming remained central to the economy of the state. This background provided an environment in which soil analyses and mineralogical investigations seemed increasingly important. Consequently, on March 12, 1872, the Ways and Means Committee of the House of Representatives reported a bill calling for a renewal of the survey. While this bill failed, it is nevertheless important, for it provided both scientific and economic objectives.[12] The funding of $10,000 was no greater than eighteen years earlier, indicating that the legislators did not intend to be extravagant. The following year the bill passed both House and Senate and was signed by Governor Preston H. Leslie on March 22, 1873.[13]

The exact manner in which N. S. Shaler came to be state geologist is not clear, but he was a native Kentuckian, a rising scientist of reputation, and was endorsed by the influential Louis Agassiz. Shaler took the oath of office on August 22, 1873, and the second survey officially began. Shaler immediately appointed Robert Peter as chemist and Albert Rogers

Crandall as geological assistant. Shaler appointed others, among whom the best known is probably Charles Joseph Norwood of Missouri.

Shaler's main difficulty was the determination of the objectives to be pursued. Kentucky was sufficiently well known by 1873 that descriptive and exploratory geology, of the type then under way in the dramatic surveys of the West, was simply not an option. A purely scientific survey was not the intent of the legislature. While scientists conducted the survey, they allowed economic objectives to dominate, as they had in the Owen survey.[14] The bill establishing the survey specified that it was to continue the Owen survey. Shaler regarded this bill as limiting, in part because of lost records. He wrote: "To my great regret, I found that almost everything had been destroyed by the burning of the offices of the State in 1869. The field notes, manuscript maps, &c., and all the valuable collections had been lost in this manner. After much search, I found in Philadelphia the copperplate of the unfinished map of Greenup, Carter, Boyd and Lawrence. . . . Dr. Peter had carefully preserved the materials in his hands at the time the survey was suspended, so that something has been saved from the work."[15]

In addition to the lack of records there was a related difficulty in determining the objectives of the initial survey. Shaler wrote: "Dr. Owen . . . left no distinct outline of his plan for the conduction of the survey, and I have not been able to see clearly what his plan was from the most attentive study of his reports."[16]

Shaler took the intent of the law seriously. Four years after beginning his work he published the "Report on the Unfinished Work of the Survey of the Commonwealth, under the Direction of Dr. David Dale Owen."[17] By that time he had given up "all hope" of recovering further records. Shaler incorporated notes of the Owen survey on maps he produced. The second survey also published a geological section drawn along the baseline of the Owen survey. The baseline itself was abandoned because it was established by outmoded methods.[18]

Although Shaler was somewhat frustrated by the ties to the Owen survey, within three months he had established his own priorities, which were followed for the remaining seven years in which the survey was under his direction. From that point on, the Shaler survey was largely independent of the Owen survey. The dated topographical work of the first survey was redone. The stratigraphical work on the coal measures was modified only slightly. The topographical and paleontological work of the Shaler survey was much more elaborate and useful than the earlier work. With the end of the Shaler survey in 1880, a colorful and creative period in Kentucky scientific history came to a close.

Notes

1. John Filson, *The Discovery, Settlement and Present State of Kentucky* (Wilmington, Del.: James Adams, 1784). This work is available in the University Microfilms, American Culture Series, no. 200, Roll 20. The Kentucky Geological Survey issued a facsimile reprint in 1929.
2. Willard Rouse Jillson, "The Advance of Geology in Kentucky During One Hundred and Thirty Four Years (1784–1918)," *Transactions of the Kentucky Academy of Sciences* 25 (1962):48. The title of Imlay's work, originally appearing in 1792, was *Topographical Description of the Western Territory of North America*. The Library of Congress has a copy of the 3rd edition (London: J. Debrett, 1797).
3. John Milton Nickles, *Geologic Literature on North America: 1785–1918*, U.S. Geological Survey Bulletin, no. 746 (Washington, D.C., 1923).
4. L. C. Robinson, "The Kentucky Geological Survey," *The Register of the Kentucky State Historical Society* 25 (January 1927):86.
5. George P. Merrill, *The First One Hundred Years of American Geology* (New Haven: Yale University Press, 1924), pp. 171, 183.
6. Robinson, "Kentucky Geological Survey," p. 86.
7. Ibid.
8. Walter Brookfield Hendrickson, *David Dale Owen: Pioneer Geologist of the Middle West* (Indianapolis: Indiana Historical Bureau, 1943), p. 116.

9. David Dale Owen et al., *Report of the Geological Survey in Kentucky* (Frankfort: J. B. Major, 1861), 4:7.

10. Ibid.

11. Ibid., p. 42.

12. *Journal of the House of Representatives of the Commonwealth of Kentucky*, regular session of December 4, 1871 (Frankfort: The Kentucky Yeoman, 1871), pp. 829, 959–61.

13. Willard Rouse Jillson, "Geological Research in Kentucky," *Kentucky Geological Survey Bulletin*, 6th ser. 15 (1923):10.

14. Nathaniel Southgate Shaler, preface to vol. 1, *Reports of Progress of the Kentucky Geological Survey*, 2nd ser. 1 (1876):v.

15. *Journal of the House*, session of December 1, 1873, p. 91.

16. Ibid.

17. *Reports of Progress of the Kentucky Geological Survey*, 3 (1877): 417–20.

18. Ibid., p. 417.

Charles Lyell's Observations on Southeastern Geology

Daniel D. Arden

Charles Lyell seems to have been amused rather than outraged when he saw a mastodon skeleton which the German adventurer Dr. Albert Koch had attempted to reconstruct using the bones from more than one animal.[1] Lyell examined Koch's exhibit in Philadelphia during one of his two extended visits to North America during the 1840s. He recognized that the tusks were turned horizontally and that other bones were improperly placed. Later he saw a skeleton of a zeuglodon, the fossil whale, unearthed in 1845 by Koch in Washington County, Alabama, and again pointed out that the great length of the "sea serpent" was the result of the bones' coming from several animals. He called Koch "a mixture of an enthusiast and an imposter, but more of the former, and amusingly ignorant."[2]

This tolerance for the foibles of others was typical of Lyell. Even when commenting critically on what he must have considered uncouth aspects of life in backwoods America, he did so without the condescending rancor evident in the observations of many cultured European visitors of the time. He found remarkably little to criticize and much to praise in America, and he did not hesitate to point out to his English readers the advantages of the American system of education and religious tolerance. His background of affluent landowner led him to the conclusion, however, that widespread suffrage was not a mark of political progress, but rather an evil that "made politics a trade."

Charles Lyell was a founder of modern historical geology and one of the most renowned scientists of the nineteenth century. Although he graduated in law from Oxford and was admitted to the bar of Lincoln's Inn, he had an early fascination with geology, which increasingly occupied his attention, to the eventual exclusion of his law practice. He was amazed to find that he could make money from his geological writings and soon established a reputation as a popular interpreter of geological ideas and observations.[3]

Lyell was at the peak of his career when he was invited in 1841 to deliver a series of lectures at the Lowell Institute in Boston. His chief work, *The Principles of Geology*, had first appeared as three volumes published between 1830 and 1833. It was under constant revision, and the twelfth edition was issued in 1876, after Lyell's death the previous year at age seventy-seven. A companion work, the *Elements of Geology*, first published in 1838, became a standard text for stratigraphy and stratigraphic paleontology. The latter was based on a series of presentations at King's College, where Lyell earned renown as a lecturer on science. He gave a series of twelve lectures in Boston to crowds of more than 3,000, necessitating the repetition of his lectures in the afternoons and evenings, and was much impressed by the decorum and varied background of the audience. The fact that he was paid three times the current rate in London may also have influenced his judgment of the success of the series.

His wife, Mary, accompanied Lyell on many of his travels, including the American visits of 1841–42 and 1845–46. On both occasions they were welcomed by the social and scientific leaders of America. Although Lyell was not knighted until 1848 and was made a baronet even later, his position as an English gentleman and scientist was well established. His credentials as president of the Geological Society and recipient of the Royal Society's medal for geology added to his prestige, especially with his American acquaintances.

The Lyells were lionized wherever they visited, and the American geologists were flattered to serve as guides for field excursions. Although he was to return to the United States for two shorter visits in 1852 and 1853, it was during the grand tours of the 1840s that the pair covered much of the eastern United States and Canada.

The first journey to the south departed from Washington in December 1841. Lyell considered Washington barbaric and

Figure 1.
Charles Lyell's American travels during the 1840s.

could not understand why it was the nation's capital or why the Smithsonian museum was being built in an area so remote from cultural centers. The pair went from Washington to Richmond, and on to Williamsburg and Jamestown, which was then deserted. Lyell described the general aspects of the Atlantic coastal plain, recognizing Cretaceous, Eocene, and Miocene rocks. Near Williamsburg he found Miocene fossils washing out of marl cliffs and collected 70 species of shells in one quarry. He says he found 147 species of mollusks as well as 13 of coral and some of barnacle from the Miocene in this region. He considered them to be Miocene on the basis of stratigraphic position above Eocene beds, affinity with European fossils, and proportion of living species (23 of 147, or 17 percent). Lyell noted that "*species* are not the same as in Europe" and concluded that oceanic separation dated at least to Miocene and that there was a warmer environment in the Miocene of Virginia than in England.

The Lyells left Williamsburg on December 23 by riverboat for Norfolk, and then went to Weldon, North Carolina, by rail. During this journey the train encountered a sleet storm and was able to proceed only by backing and running forward until momentum was lost and the wheels began to spin. They passed the Great Dismal Swamp, which Lyell described. He compared it with Carboniferous coal swamps and noted that the swamp was about twelve feet higher in elevation than the surrounding land, and that streams drained away from rather than into the swamp. It was in North Carolina that Lyell first encountered rocking chairs, which he thought a useful invention. They continued on to Wilmington by rail, then by sea to Charleston. He considered the transportation system and accommodations remarkable, especially the rail trip from Charleston to Augusta, Georgia; the 136-mile journey was made in one day.

The couple spent three days in Augusta, where they saw their first raw cotton. Lyell examined beds of white kaolin along the Savannah River and concluded they were of Eocene age. On New Year's Day 1842 they took a cotton boat forty miles downriver, recognizing the topographic expression of natural levees. They landed at Shell Bluff, where Lyell found forty species of Eocene shells, and spent some time at a plantation near Demery's Ferry. The Shell Bluff locality, abounding with large oysters (*Crassostrea gigantissima*), was a well-known attraction on the river and had been seen by many visitors, including William Bartram. T. A. Conrad, whom Lyell characterized in a letter to his father-in-law, Leonard Horner, as "the best-informed paleontologist on this side of the Atlantic," was at Shell Bluff shortly after Lyell's visit and confirmed his friend's conclusion that the beds were of Upper Eocene age.

The Lyells went thirty miles farther to Stony Bluff to examine the Eocene "burr stone." They stayed at Millhaven with a Colonel Jones, who took Lyell to Jacksonborough. He found a large number of the inhabitants afflicted with malaria, which he believed to be somehow related to boggy soil that was exposed to hot sun and affected by alternating wet and dry conditions.

The travelers continued some 100 miles to Savannah by carriage, examining the geology on the way and staying at country inns and homes. They arrived in Savannah on January 10. Lyell was directed by a Dr. Habersham to the site of a vertebrate fossil find at White Bluff Creek, twelve miles from Savannah, where he found numerous Pleistocene bones and teeth and saw evidence suggesting a recent rise in sea level.

They returned to Charleston by ship on January 13. Lyell had collected or been given 125 species of Eocene shells, of which he found seven species common to the Eocene of Europe.

The journey north was by ship from Charleston to Wilmington, and overland to Philadelphia. On the way, Lyell collected fossils from Miocene marl and Claiborne limestones, and found Cretaceous rocks at Washington, North Carolina. Michael Tuomey accompanied him on a field trip to Petersburg, Virginia, and he saw Joseph Nicollet in Washington. They reached Philadelphia on February 1, at the time of Charles Dickens' visit, and Lyell spent six weeks lecturing. Before sailing from Halifax for Liverpool on August 18, 1842, the Lyells toured the Ohio Valley and the Great Lakes region. They crossed the Alleghenies, saw the midcontinental coal fields, and visited Big Bone Lick, Kentucky. Altogether, they had been away from England for more than thirteen months.

Travel in the nineteenth century was seldom a routine affair, and few persons had either the stamina or the financial resources for extended journeys. Thus, vicarious expeditions by way of books were as close as most ever got to seeing the world. Travel books were popular, and one by a gifted writer such as Charles Lyell was sure to be successful, as indeed it was. His *Travels in North America* appeared in two volumes in 1845 and went through several printings in various formats, both in Great Britain and the United States. In addition, some dozen papers of more scientific content were published as a result of the voyage.

The success of the first visit and the opportunity for further subsidized lectures lead to another American tour by Charles and Mary Lyell. They departed from Liverpool on the *Britannia* on September 4, 1845, and after some rough weather and sightings of icebergs, reached Boston on September 19. That the Lyells spent only one hour in customs says something for their standing in official eyes.

After travels in New England, where Lyell saw glacial and marine Pleistocene beds, the two left Boston for the south on December 3, 1845. On the way they were guests of Benjamin Silliman in New Haven, took steamers to New York, and in New Jersey observed the Palisades and saw Triassic rocks. Lyell was no more favorably impressed with Washington than on his previous visit, considering it a development scheme that failed. He was entertained by Mrs. Polk at the White House and met General Scott, Captain Wilkes (of Antarctic fame), and Navy Secretary Bancroft. He examined coal mines belonging to an English company thirteen miles west of Richmond, and recognized two coal regions—the western coal fields, of Carboniferous age, having plants analogous to the European coal measures, and another province east of the Appalachians, typified by the James River coal fields, which rested on "granite" and was of Liassic age.

They departed Richmond on December 23 by rail for Wilmington, collected Tertiary fossils near there, and took a ship to Charleston, arriving on Christmas Day. From Charleston they took the S.S. *General Clinch* ("a fine steamship") to Savannah. Mary Lyell was amused when a woman, a fellow passenger from the west, told her, "Your old man is mighty eager to see you." The ship followed the inland waterway and became stranded for a few hours at low tide, which did not distress Lyell as it gave him an opportunity to observe coastal topography and vegetation. He noted changes in the coastal lands, such as encroachment of the sea at Cockspur Island, while there was deposition of coastal sands at other points.

Lyell noted that Savannah had a population of 12,000 and "seems rather stationary." He took excursions with Dr. LeConte (father of Joseph) to Skidaway Island, where they dug bones of Pleistocene mammals from cliffs at the north end of the island. These included bones of *Megatherium*, *Mylodon*, *Elephas primigeneus*, and *Mastodon giganteus*.

On December 31 the Lyells took a steamer south from Savannah to Darien, Georgia, where they were met by Hamilton Couper in a long canoe hollowed from a cypress log and rowed by six slaves. Couper was an important landowner, wielded political power in Georgia, and had earned a reputation as a careful scientific observer. He supported scientific investigations of various kinds and had contributed papers to several scientific journals. He escorted the Lyells up the Altamaha River to his principal plantation at Hopeton.

The Lyells enjoyed the hospitality at Hopeton, where they remained for ten days. During their stay they visited St. Simons Island, where they saw an Indian shell mound on the northeastern end of the island. While cruising the coastal rivers Lyell noted that shale clasts may be found in sandstone where alternating clay and sand beds are being deposited, and that rivers become turbid when adjacent forest land is cleared. He recognized coastal terraces, ten to forty feet in elevation and extending inland for twenty miles, of fluviomarine character and Pleistocene age, resting on Eocene beds. Second and third terraces he said were developed on Eocene terrain, each about seventy feet higher than the one previous. He assumed these were ancient shoreline cliffs and attributed all of the terraces to changes in land elevation rather than sea-level changes.

Lyell commented on the ill-fated Brunswick canal, an attempt to open navigation between the Turtle and Altamaha rivers, on which about $900,000 had been spent. He noted that no useful geological results were obtained except for the recovery of many Pleistocene vertebrate remains found in fluviatile clays overlying marine sands. These fossils were recovered largely through the intercession of Hamilton Couper.

From Hopeton the Lyells returned to Savannah and then went to Macon, Georgia, by rail. Mary made the trip to Macon in a regular coach while Charles went by handcar, propelled by three black men. He chose this method of transport because he could stop at cuts and look at geologic sections. Along the way he saw the limestone at Sandersville, which he recognized as Eocene in age. They reached Macon on January 15 and saw Indian mounds as well as a blockhouse, still on location, that had been used against Indian raids.

From Macon they went to Milledgeville, which in 1846 was the capital of Georgia. Dr. J. R. Cotting showed Lyell fossils he had collected and took him to localities around the town. The journey continued westward by stagecoach to Columbus, Georgia, a route that approximately followed the Fall Line. They crossed the Flint River on a ferry near Knoxville during an all-night trip. Lyell noted that the last group of 500 Indians had been sent from the area to Arkansas only a week before.

They continued from Columbus to Chehaw, Alabama, by coach, and then went by rail to Montgomery, where Lyell recognized Cretaceous strata. He decided it would save time and money to go from Montgomery by steamer down the Alabama River to Mobile, then up the Tombigbee to Tuscaloosa, because the Cretaceous terrain was so treacherous in wet weather. This itinerary provided opportunities for fossil collecting from river bluffs. An excellent collection was made at Prairie Bluff, and they stopped for several days at Claiborne. Lyell learned of some zeuglodon finds around Clarkesville, Alabama, and concluded that these were the bones of a primitive whale, which indeed they were.

From Tuscaloosa Lyell was taken to the coal fields around Birmingham and expressed the opinion that the iron, limestone, and coal of the district would ultimately be a source of mineral wealth. At Demopolis he saw fulgurites and recognized their origin as sand fused by lightning bolts. He examined the Tertiary limestone deposits at St. Stephens on the Tombigbee River, where this material is still actively quarried, and thought the huge foraminiferids, *Lepidocyclina*, were corals.

From Mobile the pair went by inland steamer to New Orleans, arriving at Mardi Gras time. Lyell spent three days with Dr. William Carpenter, a physician and naturalist of New Orleans, in the Mississippi delta area. The couple then went up the Mississippi, carefully observing topographic and geologic features of the great river. A side trip was made from

Vicksburg by rail to Jackson, Mississippi, and included other brief excursions at various points. They spent several days around New Madrid, Missouri, where evidence of the earthquakes of 1811–12 was still fresh. He concluded that the loess was alluvium deposited during an episode of downwarp and later uplifted, and compared this with terrace deposits along the Atlantic coast.

They continued upriver to Ohio, passing Cairo, Illinois, on March 29. They visited New Harmony, Indiana, and other points near the Ohio River country and returned to Boston on May 23, leaving for England on June 1, 1846.

It is likely that no other early naturalist, with the possible exception of William Bartram, had seen so much of eastern North America as Charles Lyell, and certainly no one else had described so accurately the geology and topography of the region. He made some errors in interpretation, but for the most part he was a careful observer who approached an outcrop without preconceptions.

Notes

1. Albert Koch was born in Germany in 1804 and came to the United States about 1826. He established and operated a private museum in St. Louis, Missouri, from 1836 until 1841, when the museum and most of its exhibits were sold. Koch did not part with his mastodon skeleton, however, and exhibited it in Europe, where it finally came to rest in the British Museum. A report on Koch, especially his years in Missouri, has been written by John Francis McDermott, "Dr. Koch's Wonderful Fossils," *Missouri Historical Society Bulletin*, no. 4 (1948):233–56.

2. Letter from Charles Lyell to his father-in-law, Leonard Horner, included in *Life, Letters and Journals of Sir Charles Lyell, Bart*., vol. 2, ed. Katherine M. Lyell (London: J. Murray, 1881), p. 76.

3. The most completely documented account of Lyell's life and accomplishments up to the time of his first American visit is in Leonard G. Wilson, *Charles Lyell, the years to 1841: The revolution in geology* (New Haven: Yale University Press, 1972).

Appendix

Publications by Charles Lyell relating to his travels in the Southeast:

1. "On the Tertiary Formations and Their Connexion with the Chalk in Virginia and Other Parts of the United States." *Proceedings of the Geological Society of London* 3 (1838–42):735–42. Description of various Tertiary areas of the Southeastern coastal plain, including Eocene at Shell Bluff and subdivisions of the Eocene.

2. "On the Geological Position of the *Mastodon giganteum* and Associated Fossil Remains at Bigbone Lick, Kentucky, and Other Localities in the United States and Canada." *Proceedings of the Geological Society of London* 4 (1843–45):36–39. The paper was reprinted in the *American Journal of Science* 46 (1844):320–23.

3. *Travels in North America in the Years 1841–2; with Geological Observations on the United States, Canada, and Nova Scotia*. 2 vols. New York: Wiley and Putnam, 1845. Several editions were issued by various publishers in Great Britain, United States, and Germany.

4. "Miocene Formations of Virginia and North Carolina, &c, with Appendix, on Fossil Corals, by Mr. Lonsdale." *Proceedings of the Geological Society of London* 4 (1843–45):547–63.

5. "On the White Limestone of South Carolina and Georgia, and the Eocene Strata of Other Parts of the U.S., with Appendix, on the Corals, by Mr. Lonsdale." *Proceedings of the Geological Society of London* 4 (1843–45):563–76.

6. "Notes on the Cretaceous Strata of New Jersey and Other Parts of the United States Bordering the Atlantic." *Quarterly Journal of the Geological Society of London* 1 (1845):55–60.

7. "Observations on the White Limestone and Other Eocene or Older Tertiary Formations of Virginia, Carolina, and Georgia." *Quarterly Journal of the Geological Society of London* 1 (1845):429–42.

8. Extracts of letters from Charles Lyell to Benjamin Silliman:

"Claiborne, Ala., dated Feb. 4, 1846, on the *Zeuglodon* fossils and Eocene limestone of Washington and Clarke Counties, Ala." *American Journal of Science*, 2nd ser. 1 (1846):313–15.

"Mobile, Ala., Feb. 19, 1846. Coal field of Tuscaloosa, Alabama." *American Journal of Science*, 2nd ser. 1 (1846):371–76.

9. "Observations on the Fossil Plants, of the Coal Field of Tuscaloosa, Alabama; with a Description of Some Species by C. T. F. Bunbury." *American Journal of Science*, 2nd ser. 2 (1846):228–33.

10. "On the Newer Deposits of the Southern States of North America." *Quarterly Journal of the Geological Society of London* 2 (1846):405–10.

11. "On the Delta and Alluvial Deposits of the Mississippi, and Other Points in the Geology of North America, Observed in the Years 1845, 1846 (Abstract of a discourse before the British Association, Sept. 14, 1846, from the Athenaeum for September 26)." *American Journal of Science*, 2nd ser. 3 (1847):34–39.

12. "On the Alleged Coexistence of Man and the Megatherium (addressed as a letter to the Editor of the *Times*, and published in the *Times* [London] of Dec. 8, 1846)." *American Journal of Science*, 2nd ser. 3 (1847):267–69. Lyell discounts the theory that a fragment of human pelvic bone found near Natchez, Mississippi, is from the same horizon as bones of *Megatherium*.

13. "Geological Society of London, April 14th, 1847. A paper was read, entitled 'On the Structure and Probable Age of the Coal Field of the James River, Near Richmond, Virginia.'" *American Journal of Science*, 2nd ser. 4 (1847):113–15.

14. "On the Relative Age and Position of the So-Called Nummulite Limestone of Alabama." *American Journal of Science*, 2nd ser. 4 (1847): 186–91.

15. *A Second Visit to the United States of North America*. 2 vols. New York: Harper, 1849. A second edition, revised and corrected, was published by John Murray, London, in 1850.

16. "On the Nummulite Limestone of Alabama (*Proc. Geol. Soc.*, Jan. 9, 1847)." *American Journal of Science*, 2nd ser. 7 (1849):114.

The Second Geological Career of Ebenezer Emmons: Success and Failure in the Southern States, 1851–1860

Markes E. Johnson

On March 7, 1851, Ebenezer Emmons took the stand as a key witness for the prosecution in a libel case being tried in the Albany courtroom of Judge A. L. Parker. Directly at stake was the accuracy of a geological chart showing the general succession of fossiliferous strata in North America. An event unparalleled in American geology, the drama enacted in a court of law poised the geological career of the prosecution's expert witness upon the veracity of a scientific interpretation. Every scientific age has its own preoccupations. During the early to mid-1800s geologists were absorbed with the definition of a geological time scale and the recognition of intercontinental rock formations of equivalent age. With the present emphasis on plate tectonics, it is natural for the personalities and controversies of a different century to fade from attention. That few geologists today know the name of Ebenezer Emmons is hardly surprising. Yet the conflicts involved in the ostracism of this man from the American scientific community exemplify the dangers of rigid conformity in the exercise of science during any age.

Only the bare details of Emmons' personal life were recorded by contemporary biographers.[1] Born in 1799, he was a native of western Massachusetts and was educated at Williams College. Before joining the Williams faculty in 1828,

Figure 1.
EBENEZER EMMONS (1799–1863). Emmons was founder of the Taconic System, State Geologist of New York, State Agriculturist of New York, and State Geologist of North Carolina.

initially as a lecturer in chemistry, Emmons was associated with the Rensselaer School in Troy, New York, and was also licensed to practice medicine. His ties with Williams College were unsevered for his remaining thirty-five years; he was professor of geology and mineralogy until his death. Emmons held concurrent academic appointments, however, at the Rensselaer School during the 1830s and at the Albany Medical College during the 1840s. His most important accomplishments were achieved as state geologist of New York's Adirondack region (1836–41), state agriculturist of New York (1843–51), and state geologist of North Carolina (1851–60).

Two aspects of Emmons' geological career in the northern states were extraordinary: he was a leader in using an American terminology for Paleozoic rocks, specifically his Champlain, Ontario, Helderberg, and Erie groups of the so-called New York System, and he adamantly argued that an older Taconic System of sedimentary rocks lay below the New York System, distinguished by a "primordial" fauna. This Taconic System he believed to be a rock unit of major significance, potentially recognizable on a worldwide basis.

While Emmons is now generally regarded as the first to build an American standard for reference by American geologists, his Taconic System became the focus of immediate controversy and derision among his peers.[2] The 1851 court case of *Foster* v. *Agassiz* brought Emmons to a humiliating confrontation with his opponents over the Taconic question. When the judgment went against Emmons, his retirement from geology at age fifty-two must have been a foregone conclusion to his many adversaries. Within a matter of months, however, Ebenezer Emmons was named state geologist of North Carolina. He not only set out on a fresh investigation of the Taconic System but also proved his versatility by making important contributions that elicited controversy on other fronts, as well. The purpose of this paper is to examine the success and failure of Ebenezer Emmons' second geological career in the South, which spanned nearly a decade of vigorous activity.

Turning Point

Were it not for the charges of libel pressed by James T. Foster against Louis Agassiz and James Hall, with the resulting court trial of 1851, Ebenezer Emmons might have been content to stay indefinitely at his post as state agriculturist of New York. Clearly his venture into southern geology came as a sudden decision at a critical juncture in time.

A detailed account of the unfortunate chain of events stemming from the appearance of *Foster's Complete Geological Chart* in 1849 is given by James Hall's biographer.[3] Intended for use in the secondary schools, the original edition of the Foster chart contained so many inaccuracies that Hall, state paleontologist of New York, was outraged. He enlisted the help of Louis Agassiz, Harvard University's professor of geology, in denouncing the author in the public press. Foster subsequently sued Hall and Agassiz separately for libel damages in the amount of $40,000 and $20,000, respectively. While the suits awaited trial, Foster brought out an amended edition of his chart with the notice that it had been "corrected and approved" by Professor Ebenezer Emmons. Among the substantial alterations made in the chart was the inclusion of Emmons' Taconic System.

Negatively influenced by the prevailing weight of opinion in the parallel battle among European geologists over the admissibility of Sedgwick's Cambrian System, Hall was prepared to cross Emmons at every chance. When the *Foster* v. *Agassiz* case came to trial, Hall knew Emmons would be called as a witness. Hall very carefully schooled Agassiz's lawyers. They, in turn, introduced an impressive stream of expert witnesses who downgraded Emmons' work. It was the perfect opportunity to demolish the Taconic System in court, and it worked. Foster lost his libel case and the companion case of *Foster* v. *Hall* never even came to trial. Taking advantage of the opportunity, James Hall immediately published a commercially successful geological chart of his own design.

Comparable in part to the Taconic System, the Cambrian System proposed by Sedgwick in 1836 eventually gained acceptance by American geologists. Few geologists today have ever heard of the Taconic System, much less that Ebenezer Emmons legitimately based part of his system on paleontological evidence well before the discovery of index fossils in the type Cambrian rocks of northern Wales. It was the lack of paleontological evidence that as late as 1851 made many European geologists doubtful of the Cambrian System. This skepticism was shared in the extreme by Hall, who rejected Emmons' claim to a truly "primordial" fauna. Orchestrated chiefly by Hall, the attack on the Taconic System in 1851 was expected to silence Emmons. A close friend, Albert Hopkins, recalled Emmons' stern resolve to vindicate himself: "The Doctor thought he could see, in the course pursued by his opponents, a striking analogy to that pursued by the old dogmatists, who refused to look into Galileo's telescope, for fear they should see something. I told the Doctor, on one occasion, that all the authorities were combining against him. He answered, in his dry way, 'I shall floor them yet.' "[4] Clearly it was a determined Ebenezer Emmons who accepted the position of state geologist in North Carolina and sought a change of fortune in the South.

A Matter of Public Service

That Emmons had an earlier history of manipulating his official duties to his own purpose cannot be denied. He unabashedly used his position as state agriculturist of New York to promote the geological welfare of the Taconic System. The first volume of the *Agriculture of New York*, for example, contains a sixty-seven-page section on the Taconic System, reprinted in full from a previous work, privately published and poorly circulated.[5] While Emmons undoubtedly had expectations of triumphantly extending his Taconic System to the South, he was also bound by a strong sense of public service. This sense of duty had been instilled in him by his old mentor, Amos Eaton, first at Williams College and later at the Rensselaer School.

Eaton firmly believed that science should reach out to the common people for the good of all. As his favorite student at Williams College, the young Emmons was chosen to assist his teacher on lecture tours throughout the Berkshire towns of western Massachusetts.[6] Later, the Rensselaer School was expressly founded to spread the "application of experimental chemistry, philosophy and natural history to agriculture, domestic economy [and] the arts and manufactures" for the benefit of the "sons and daughters of farmers and mechanics."[7] The Rensselaer School eventually produced seven field geologists who among themselves led nearly a quarter of the fifty-six state and federal surveys conducted between 1830 and 1860.[8] The published reports of these Rensselaer geologists generally were not written in the usual esoteric style but typically were forward in urging local populations to take full advantage of the natural resources being uncovered in their land. The same approach is abundantly evident in Emmons' North Carolina reports.

The second chapter of Emmons' 1856 *Geological Report of the Midland Counties of North Carolina* is devoted to an evaluation of river systems suitable for manufacturing sites. His discussion of the Catawba River and its tributaries makes reference to mill towns in his native Massachusetts to illustrate the potential for industrial development.

> The south or Little Catawba, though carrying much less water than the Great Catawba, may be regarded as upon the whole, the most important manufacturing river. . . . To form an estimate of the capacity of this branch of the Catawba, I may compare it with a well-known stream, the Hoosick river, in Berkshire, Massachusetts. . . . The Hoosick at Adams, is about half the size of the Little Catawba; yet it furnishes a power equivalent for turning 500,000 spindles, in a space of about twelve miles; and at the same time, there is no single location which can compare with the High Shoals of the Little Catawba. Iron, calico, satinetts and woollens are manufactured, giving employment to between 3,000 and 4,000 individuals, and making an important market for this part of the country. The Little Catawba can furnish twice the power in the same distance, and employ 10,000 persons, and create thereby a home market for the produce of all of this part of the country.[9]

Emmons makes no attempt to hide his astonishment at the lack of diversification from the agriculturally based economy in the South.

> To one who has been familiar with the disadvantages of a New England soil and climate, and who has witnessed the disappearance of formidable obstacles in prosecuting extensive public or private works by the force of the will, it will not seem strange that he should look with surprise on the little progress which a people under a better sun, with a better soil, and numerous natural advantages to encourage, have made in the last quarter of a century. When, however, a single public or private enterprise has been prosecuted to a successful termination, confidence is increased and timidity diminished. Every successful enterprise multiplies the friends of enterprise, and the results are being seen in the growth of villages, the erection of a better class of dwellings, and the circulation of money.[10]

Ironically, the northern mill towns are now floundering in an energy-hungry climate. Although the mill towns never came to the Catawba River, the present state of industrial prosperity in the Sun Belt is what Emmons once envisioned.

Other chapters in the 1856 report deal with the geology of North Carolina's mineral resources, such as coal from the Dan River and Deep River fields, as well as iron ore and more precious minerals from the midland counties. Ever on the lookout for areas of potential development, Emmons' text is filled with encouragements for local enterprise. A later report deals creditably with the agriculture of North Carolina's eastern counties, although a third of the text is taken up with paleontological descriptions of Cenozoic fossils.[11] Today Emmons' original descriptions of new vertebrate species are especially valuable. It is questionable, however, whether this was what the people of North Carolina expected in a state agricultural report.

Extension of the Taconic System

Exploration in the Southern States.

During the first five years of his southern period, Emmons traveled widely through eastern Tennessee and the Blue Ridge and Piedmont regions of North Carolina and Virginia. Finally, with the publication of his textbook, *American Geology*, in 1855, he was ready to revive the Taconic System.[12] Part two of the book contains 122 pages devoted to this task. Emmons divided the Taconic System into lower and upper parts, while confidently making three main arguments. The first is that the complete Taconic System of sedimentary rocks lies between "primitive" basement rocks and the richly fossiliferous strata of the "lower Silurian System." Mention of the New York System is dropped in favor of the "lower Silurian," or what today we would call the Ordovician System. Emmons also draws a distinction between Lower Taconic rocks, composed mostly of weathered volcanic sediments, and Upper Taconic rocks, containing far fewer volcanic products.

The second point is that Lower Taconic rocks are devoid of fossils, while Upper Taconic rocks possess a meager "primordial" fauna, fully distinct from "lower Silurian" species. Emmons places as much emphasis on the disparity in diversity between Taconic and "lower Silurian" fossils as he does on their disparity in kind.[13]

A third claim asserts that the Taconic System retains the same stratigraphic sequence in the southern states as in the northern states. Furthermore, the succession of Taconic lithologies is held to be incomparable to that of the "lower Silurian" strata. Emmons asserts that metamorphism might change the appearance of individual rock layers, but could not alter their original and unique stratigraphic arrangement.[14]

Few geologists later to study the complexities of the southern Appalachians have granted Emmons much recognition for his work. Yet he capably separated the western flank of the Blue Ridge into two major rock groups, based on river exposures studied between Harper's Ferry in what was then northern Virginia and the southwestern corner of North Carolina (Figure 2). Strata now assigned to the Ocoee Group were considered by Emmons to form the lower part of the Taconic System.[15] Strata to the west, now comprising at least the Chilhowee Group, were defined by Emmons as belonging to the Upper Taconic System.[16] By today's standards, Emmons was working with Lower Cambrian rocks (i.e., the Upper Taconic System) and Pre-Cambrian rocks (i.e., the Lower Taconic System). He also identified rocks on the Piedmont as typical of the Lower Taconic System.[17] The metasediments he examined between Wilkesboro and Raleigh would now be classed in the Charlotte and Carolina slate belts.

In October of 1856 Emmons presented his *Geological Report of the Midland Counties of North Carolina* to Governor Thomas Bragg, almost a year after sending *American Geology* to press. Several chapters of the 1856 report elabo-

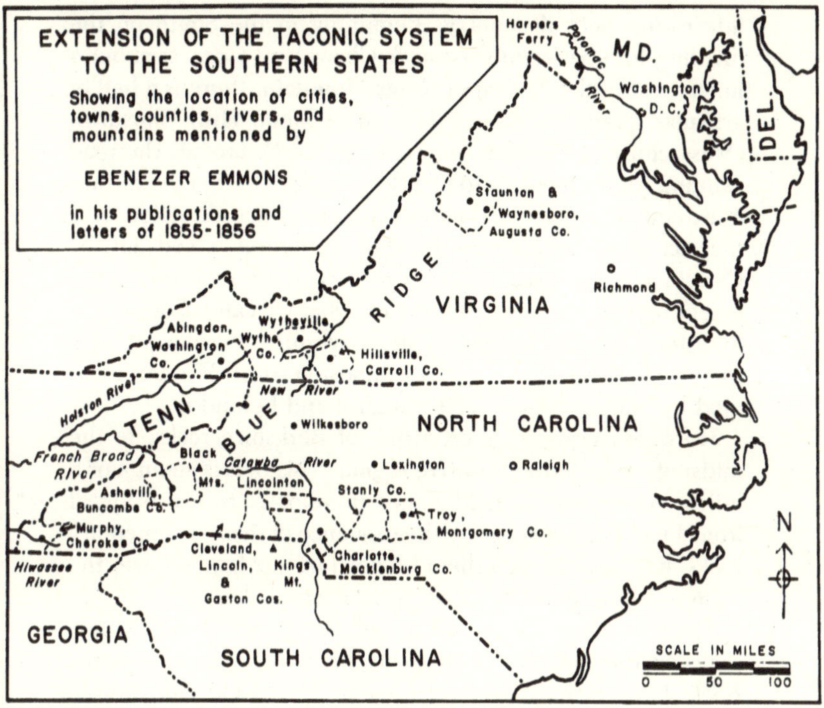

Figure 2.
TACONIC LOCALITIES. Map of eastern Tennessee, North Carolina, and Virginia, showing areas where Ebenezer Emmons studied the Taconic System.

rate on the Taconic geology of the Piedmont. A letter written on December 6, 1855, to his former teacher Chester Dewey shows how much Emmons' Taconic research dominated his mind and how closely he thought he could see resemblances to the Taconic rocks of the Berkshire Hills in Massachusetts.

> If you will take the map of North Carolina & turn to Gaston, Lincoln and Catawba Cos., I will direct your attention to one

interesting belt of country. Beginning in the South on the dividing line between Gaston & Cleaveland Cos., & about 4 miles west of the N west of Kings Mountain, there is a belt of granular quartz precisely like that of Stone Hill, Wmstown [Massachusetts]. It runs about N 20° 25° E, crosses the road running from Lincolton to Charlotte 5 miles east of the former & extends to the great Catawba [River] in that direction, it is there covered; South from this starting point it extends into S. Carolina to the Broad River which is as far as I know, the extent of it. It is accompanied with what we call talcose & the limestone is very similar to the Berkshire marbles. It dips some west-ward, but very thick almost vertical. It is accompanied by magnetic iron ore through it and behind it.

Here you perceive then a belt of Berkshire rocks, in the midst of a decidedly primitive region. This belt is about 4 or 5 miles wide, contains gold mines in veins & brown oxide of iron. I had been suspecting the existence of these sedimentary rocks, but had not seen them till now, having hopped over the country at points where they were concealed.

Their existence makes out a complete series of the Taconic rocks on the east side of the Blue Ridge. The upper Taconic Rocks I recently passed over in going from Gold Hill, Rowan Co., to the Yardkin, Mountgomery Co., —But the upper and lower divisions are separated by pyrocrystalline rocks. I have traced the quartz rocks from near the South Carolina line almost continuous to the Catawba river, a distance of 60 miles at least.——You preceive that I am perservering in working out this important system. I am getting additional facts every day.[18]

The discovery Emmons made would today be considered part of the Kings Mountain belt. This belt, together with the Charlotte and Carolina slate belts, are regional zones of different metamorphic grades imposed on sediments and volcanics of mixed Paleozoic age.[19] Here Emmons encountered the same difficulty he faced with the type Taconic rocks of the north, by insisting that their age in present terms was exclusively Cambrian and Pre-Cambrian.

The last published summary of his Taconic researches along the full range of the Appalachians was made by Emmons in his *Manual of Geology*.

> The range of country through which this series passes is the western face of the Green Mountains, extending from Canada to Georgia. These rocks cross the Blue Ridge at Harper's Ferry, or in its vicinity. They occupy also a range of hills or mountains, east of Wytheville, Va., extending westward to the immediate vicinity of the Queen's Knob, where they face unproductive coal measures. The series also passes along the eastern flanks of the Green Mountains and Blue Ridge, passing through the eastern part of Virginia and North Carolina, and onwards into Alabama.[20]

In geological retrospect, much of the stratigraphic detail Emmons tried to correlate between the Taconic rocks of the North and South is simply lacking, particularly with regard to the Piedmont. In general terms, however, his work on the western flank of the Blue Ridge was quite successful. The age relationships he postulated are crudely correct by today's standards. In some places their correspondence to rocks in the North is striking. Recently Rodgers confirmed that part of the Chilhowee Group in northern Virginia is "entirely comparable to the basal Cambrian clastic sequence on the west flank of the Green Mountain anticlinorium in Vermont."[21]

Fossil Evidence in the Southern States.

Where possible, Emmons tried to defend the Taconic System on the basis of paleontologic evidence. Unhappily, he was far less successful at this task in the South than he had been in the North. A friend, Dr. Asa Fitch, discovered trilobites within the "black slate" of Bald Mountain in Washington County, New York, which Emmons subsequently named, figured, and described as Taconic fossils in a pamphlet of 1844.[22] Considered valid taxa today, *Elliptocephala asaphoides* (Figure 3a) and *Atops trilineatus* (Figure 3b, c) are assigned to the Lower Cambrian; these and other trilobites assigned to the Middle and Upper Cambrian now are well known in the Taconic region of eastern New York.[23]

At the time of their discovery, James Hall was convinced that *Elliptocephala* and *Atops* were misnamed "lower Silurian" forms. Emmons graciously offered the explanation that these trilobites "were first discovered in fragments, and hence there were reasons for difference of opinion respecting them."[24] Obligingly, he found a superior collecting locality near Bald Mountain and produced a well-preserved specimen of *Atops* (Figure 3d), which he used to illustrate his 1855 *American Geology* and 1860 *Manual of Geology*. Meanwhile, additional trilobite faunas were being discovered by other geologists in Vermont and Quebec. The sharp debate that accompanied all such northern discoveries is well summarized by Merrill.[25] For almost seventeen years, until the influential French paleontologist Joachim Barrande took his side in 1860, Ebenezer Emmons stood by himself in proclaiming an American "primordial" fauna.

Because he accepted the notion that the Taconic Period was one of limited life, Emmons was not especially frustrated by his inability to find Taconic fossils in the South. He did name, figure, and describe a single Taconic trilobite from Augusta County, Virginia, in his 1855 *American Geology*. Emmons' *Microdiscus quadricostatus* (Fig. 3e) seems to be taxonomically misplaced today. The whereabouts of the original specimen is unknown, but the figure is "a fairly decent rendition of a trinucleid trilobite" allied to the post-Cambrian genus *Cryptolithus*.[26] Apparently, Emmons stumbled onto fossiliferous strata of Ordovician age in his Waynesborro-to-Staunton transect of Virginia's Blue Ridge (Figure 2).

Later, similar misinterpretations were made. Prominently featured in his 1856 *Geological Report of the Midland Counties of North Carolina* is the discovery of fossil corals in the

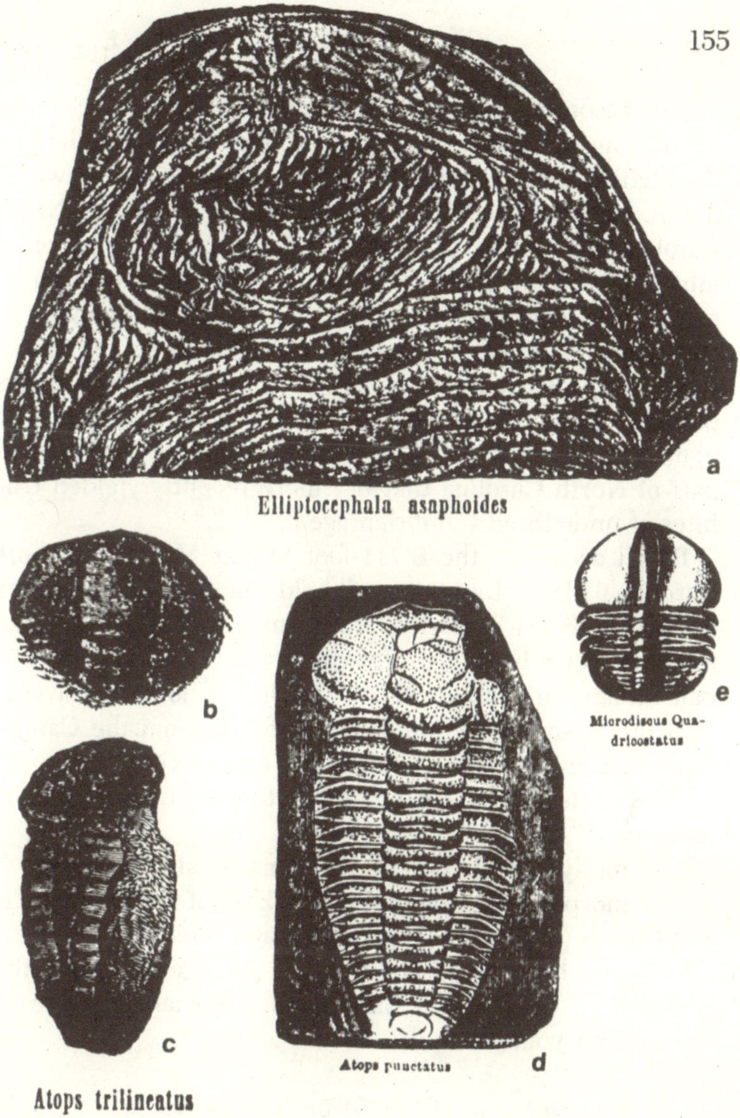

Figure 3.

UPPER TACONIC TRILOBITES. *Elliptocephala asaphoides* and *Atops trilineatus* were first described by Emmons from the northern Taconic in 1844. *Microdiscus quadricostatus* was described from the southern Taconic in 1855. *Atops punctatus* came from the northern Taconic at about the same time. All of Emmons' trilobite descriptions were foci of controversy.

Lower Taconic: "It is sometimes amusing to see the claim set up with an obvious feeling of pride, that North-Carolina has the highest peak east of the Rocky mountain range. It will no doubt be amusing to others, should I claim for North-Carolina, the honor of being the birth place of the oldest inhabitants of this globe. The fossils of Montgomery county, to which I refer, and to which I have given the family name, *Palaeotrochis*, or old messenger, are quite likely to prove in reality, the oldest representatives of the mysterious principles [of] life." Two species of *Palaeotrochis* (Figure 4) were figured and described.[27] The original collecting locality is in a part of North Carolina that has more recently yielded trilobites of undoubted Cambrian age.[28]

By all accounts, the 6,711-foot Mount Mitchell of North Carolina's Black Mountains still holds the title of highest peak in eastern North America. Claim to the title of the oldest fossil has since been lost. Yet even today, Emmons' claim to paleontological fame for North Carolina is not as extravagant as it might sound. Wormlike trace fossils from the Carolina slate belt in North Carolina have only recently been described as the "oldest well-dated metazoan fossils from North America."[29]

Unhappily, the larger of Emmons' fossil corals may be pseudomorphs of inorganic origin.[30] Even if the smaller of his fossil corals (Figure 4a–c) are legitimate, their reputed association with bryozoans would imply an age no older than Ordovician. Once again, Emmons may mistakenly have collected from younger beds.

Outcome of the Taconic Controversy.

Rodgers divides the Taconic controversy into three historical phases: (1) a Wernerian phase (1842–60), (2) a faunal phase (1860–88), and (3) a later tectonic phase.[31] The year 1842 was the year Emmons' *Geology of New York*, part 2, appeared,

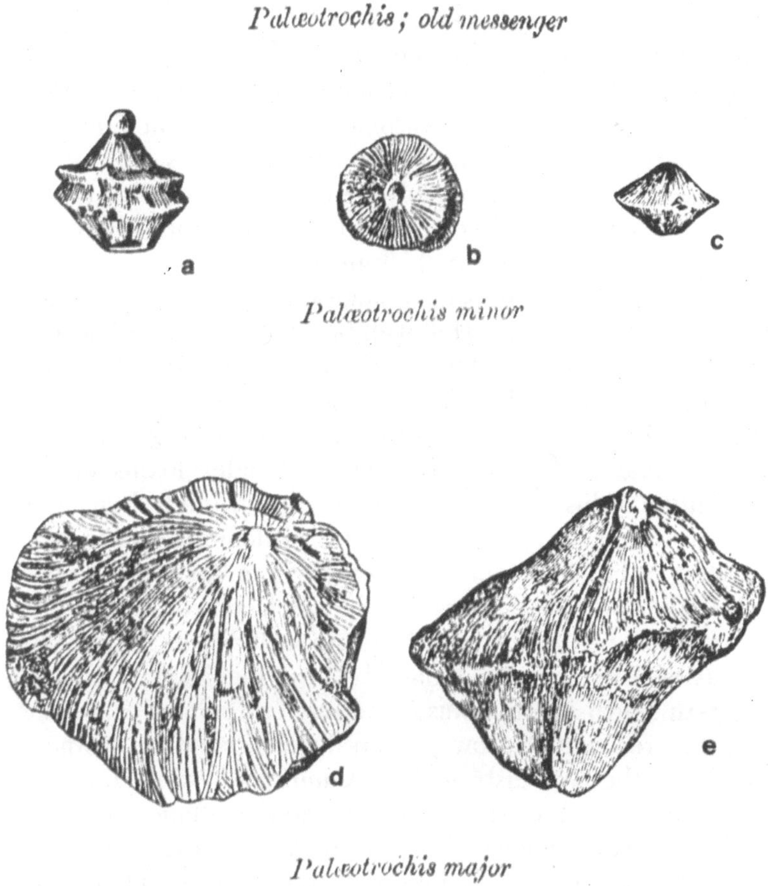

Figure 4.
LOWER TACONIC CORALS. *Palaeotrochis major* and *Palaeotrochis minor* were first described by Emmons in 1856 and identified as coming from the Lower Taconic. They were, of course, quite controversial. The larger is now interpreted as inorganic and the smaller as Ordovician.

with the first major statement on the Taconic System. Emmons was by no means a Wernerian geologist, but he did pay overriding attention to lithology as the principal factor in the establishment of a detailed Taconic stratigraphy and its correlation.[32] A penchant for including metamorphosed strata of diverse ages within the Taconic System brought Emmons into sharp conflict on purely geological grounds with other Appalachian geologists, notably Henry D. Rogers, state geologist of New Jersey and Pennsylvania, and his brother William B. Rogers, state geologist of Virginia.

Sharp conflicts on paleontological grounds were fought with the likes of James Hall well before 1860, when Joachim Barrande entered the fray. Perhaps it is unjust to exclude Emmons from the faunal phase of the controversy simply because he had so few fossils with which to work. Emmons clearly understood the importance of index fossils in the definition of his New York System. If available to him, they surely would have been put to use in defining the Taconic System. The faunal phase justifiably can be said to have closed when Walcott showed to the satisfaction of all that both Cambrian and Ordovician faunas are included in the Taconic rocks of Emmons' definition.[33] Thus, the Taconic area of eastern North America was eliminated as a serious contender for the recognition now universally conceded to northern Wales as the type-district of the Cambrian System. Interestingly enough, the Taconic controversy continues today in New England, where the plate-tectonics model of global geology suggests new solutions to the many problems of a structurally complex region.

Newark Group Sandstones in the South

Another Conflict over Geologic Age and Correlation.

Triassic beds of the Newark Group occur in faulted basins stretching from Connecticut to the Southern Appalachians

(Figure 5). The Dan River deposits of Virginia and North Carolina, and the Deep River deposits of North Carolina and South Carolina were of tremendous interest to Emmons. The final third of his 1856 *Geological Report of the Midland Counties of North Carolina* and all of his 1857 *American Geology* (part 6) are devoted to the stratigraphy and paleontology of the Dan River and Deep River coal fields. Previous studies had been made in the Connecticut River valley by Edward Hitchcock, state geologist of Massachusetts, and in New Jersey, Pennsylvania, and Virginia by the Rogers

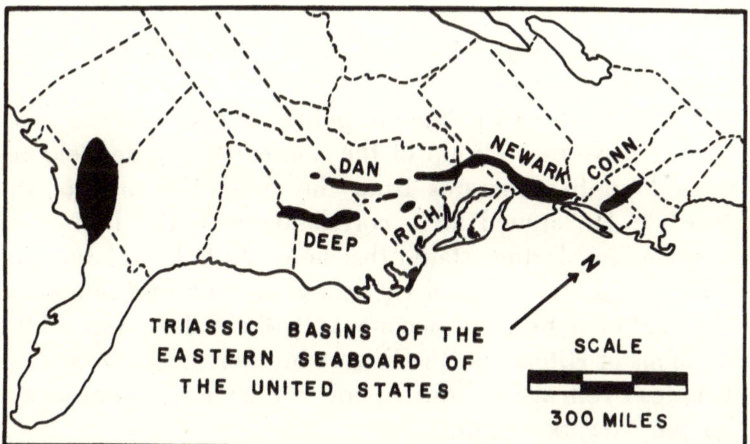

Figure 5.
TRIASSIC BASINS. Map of the eastern seaboard of the United States, showing the location of Triassic basins. Ebenezer Emmons studied the Dan River and Deep River basins in Virginia and North Carolina and was the first to declare their Triassic age.

brothers. Emmons' extension of the Newark Group to southern Virginia and North Carolina, based on extensive fossil evidence, demonstrates that he was a versatile geologist not limited by the problems of Taconic geology.

The high diversity of organisms preserved in the rocks of the Newark Group still attracts attention. A recent article on the Dan River basin lists an impressive array of reptiles, fishes, insects, crustaceans, and vascular plants dated as Late Triassic.[34] During the 1850s, however, there was wide disagreement as to the exact age of the Newark Group and its correlation with European rocks. In his 1856 report, Emmons separated the Newark Group into two subunits: an upper division, which he correlated with the Upper Triassic Keuper beds of Europe, and a lower division, which he identified as belonging to the Permian System. This distinction was drawn on the occurrence of key plant fossils and labyrinthodont amphibian remains in the upper layers, as compared to thecodont reptile remains in the lower layers. These latter fossils he mistakenly believed were restricted to rocks of Permian age.

The same year that Emmons published his report on fossils from the Newark Group of the South, W. C. Redfield reported on fish remains from the Newark Group of the North.[35] He argued for a correlation with the Jurassic of Europe, and further stated that he could find no taxonomic difference between the fossil fishes collected by Emmons in the South and his own specimens. Redfield also reviewed the opinions of Hitchcock, the Rogers brothers, Agassiz, and Sir Charles Lyell, all of whom supported the idea of a Jurassic age for the Newark Group.

Emmons clearly recognized the massive opposition he faced: "It is with considerable diffidence that I have proposed the separation of the upper sandstone from the sediments beneath, and it is only after a careful examination, and mature deliberation, that I have ventured to adopt the plan, and publish it under the circumstances I am placed in, knowing very well that it is not in accordance to the published opinions of our most learned geologists. I find, however, that the phenomena do not admit of any other interpretation, in light of facts now known."[36] Another conflict was developing along much the same lines as the Taconic controversy. Emmons was alone again on one side and many of his former opponents were on the other.

In the following year, Chester Dewey published a review of Emmons' report, which introduced the verdict of Professor O. Heer, of the Federal Polytechnic School in Zurich.[37] The Swiss professor had been asked to examine Emmons' report and many of Emmons' plant fossils. Heer concluded that several "are characteristic of the Keuper or Marnes Iresees of Germany, of France, and of Switzerland; some others are like species that are found in Europe in the Keuper and the inferior Lias, but are specifically different. There is no species really Oolitic (Jurassic) in Virginia and North Carolina." To this article, James D. Dana appended a short note stating that the Newark Group should be considered partly Jurassic and partly Triassic in age.

Shortly thereafter, Emmons issued part six of his *American Geology,* in which he restated his views on the Triassic and Permian ages of the Newark Group in the South. Included in this volume is extensive coverage of the Newark Group's flora and fauna, the most spectacular of which are the skull bones of the labyrinthodont *Dictyocephalus elegans* (Figure 6a) and the jaw of the small "mammal" *Dromatherium silvestre* (Figure 6b). Emmons was extremely proud of his "mammal" discovery and he described it himself. Because it was associated with thecodont remains in the lower layers of the Newark Group, he proclaimed *Dromatherium* to be a Permian mammal. Curiously, the source of Emmons' reaffirmed belief in the restriction of reptilian thecodont fossils to rocks of Permian age is implied to be the anonymously authored book *Vestiges of the Natural History of Creation*.[38] The concept of the "progression of types" is explicitly cited by Emmons as the fundamental phenomenon behind the limitation in range of index fossils. Yet it is difficult to judge what Emmons actually thought about this scheme of evolution, since his discovery of a Permian "mammal" would refute the timing and order of the "progression of types" observed by pre-Darwinian evolutionists.

Real and Supposed Motives for Opposition.

From the printed literature, it is not easy to assess the extent to which Emmons' "mammal" became the center of debate. Even before his formal announcement of *Dromatherium*, Emmons won a partial victory with Heer's verdict that the Newark Group in the South contains plants similar to those of the Triassic Keuper beds in Europe. Letters between Emmons and the Hopkins brothers at Williams College suggest that he was sensitive about the treatment he had received in this dispute:

> Today I received another letter from Prof. Marcou of Zurich saying that Sir Charles Lyell had spent 3 weeks in Zurich, & after examining the question respecting these rocks (Connecticut river Sandstone, Virginia, N. Carolina) had come to the conclusion that the doctrine of my report & Part 6th of my American Geology was the true doctrine. . . .
>
> The rock which contains that beautiful little jaw, is now regarded as either Permian or at least no newer than . . . lower Trias; & this mammal is, as I have maintained, the oldest or most ancient mammal of the globe up to this time. So I am fully sustained, which is a matter of some interest to me, in as much as I was abused in Silliman's Journal for maintaining this doctrine.[39]

In another letter, Emmons expressed the opinion that "Mr. Lyell has more magnanimity than the rest," referring directly

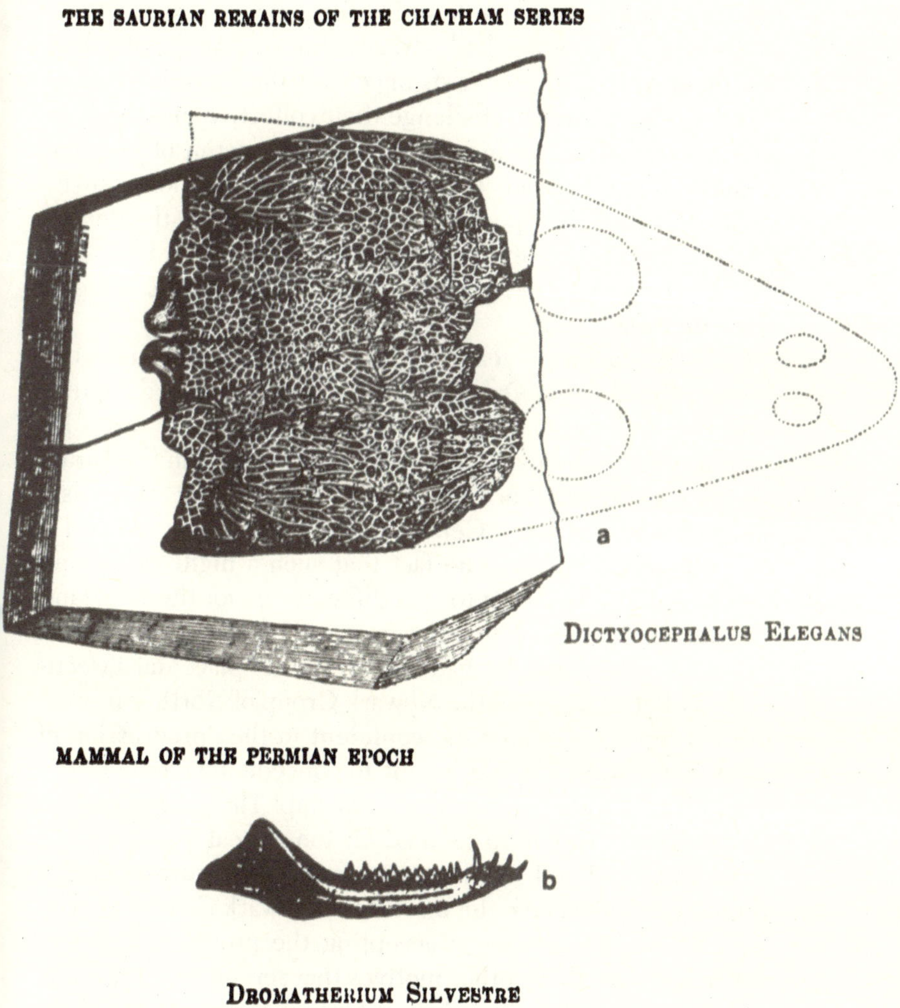

Figure 6.
EMMONS' VERTEBRATES. Emmons' description, in 1857, of fossils that he interpreted as a Triassic labyrinthodont amphibian and a Permian mammal created a global reaction. The fossils remain enigmatic today.

to H. D. Rogers, Hall, Redfield, and Dana.[40] Why did Emmons face so much opposition at home? Why was Sir Charles Lyell so quick to lend his support?

Ebenezer Emmons had already bucked an established coterie of American geologists once over the Taconic System. When he came back to challenge their collective judgment on still another issue, he may have been the victim of extreme prejudice. At least, this is probably the way Emmons saw it. If the "mammal" jaw played a more central role in the debate over the age of the Newark Group than hinted at in Emmons' letters, there may have been other reasons for the general opposition he faced.

Only thirty-three years earlier, William Buckland, the reader of geology at Oxford, enjoyed an untarnished reputation, but he met an almost solid front of criticism when he announced the discovery of mammalian remains in the Jurassic Stonesfield Slate of Oxfordshire.[41] The oldest previously known mammals were Cenozoic and most geologists had a difficult time accepting the fact that such a highly advanced group could be so old. It took eighteen years for the Buckland dispute to be settled to everyone's satisfaction. Next came Emmons' claim that he had discovered a "placental insectivore" of Permian age in the Newark Group of North Carolina. For conservative geologists, confident in the "progression of types," it required a major shift to concede a Triassic rather than a Jurassic age for the Newark Group. The suggestion of a Permian age must have seemed far too radical a position.

Sir Charles Lyell was the most important figure to accept Emmons' claim of an older age for the Newark Group. Was he merely "magnanimous" in accepting the proofs of new evidence, or were there other motives that served his purpose in making such a shift in opinion? One of the great questions geologists debated during the early 1800s was whether the physical and biological development of our planet was progressional by nature or essentially in a steady state.[42] Lyell's views as a physical uniformitarian are well known. He recognized no manifestation of a directional change in the earth's environment through time. Perhaps less appreciated today is Lyell's pre-Darwinian insistence that the organic world conformed to the same steady-state laws of the physical world. In the sixth edition of his famous textbook, Lyell contended that "we must not, however, too hastily infer from the absence of fossil bones of mammalia in the older rocks, that the highest class of vertebrated animals did not exist in remoter ages."[43] If mammals had lived in the Jurassic Period, why not in the Permian Period? While he never questioned the mammalian identity of *Dromatherium*, Lyell did settle eventually on a Late Triassic age for the lower division of the Newark Group. The North Carolina fossil and its ascribed Triassic age are cited in the 1873 edition of *Principles of Geology*, but with the admission that the author's belief in a steady-state organic world had been shaken.[44] By this time, Darwin was beginning to have an influence on Lyell.

Outcome of the Newark Group Controversy.

Deposition of the Newark Group is today accepted as an event of Late Triassic age. In subsequent years Emmons' "placental insectivore" has not fared as well. Originally he had collected three jaws, all of which were attributed to *Dromatherium silvestre*. Henry Fairfield Osborn redescribed them and split off a new mammalian genus and species: *Microconodon tenuirostris*. Simpson concluded that *Dromatherium* and *Microconodon* are not mammals, but cynodonts, mammallike therapsid reptiles. A mammalian affinity for *Microconodon* was later upheld by Kühne. The most recent compendium on Mesozoic mammals reports that no new discoveries have been made to date, but suggests that a more complete specimen of *Microconodon* would be necessary to prove Kühne's contention.[45] Whether they represent mammallike reptiles or true early mammals, their discovery was highly significant.

War Fever and the End of a Career

Emmons' geological career in the South remained active until 1860, the year his *Manual of Geology* appeared. Letters sent north to friends show that he was very concerned over the secessionist movement, but determined to remain at his post with the geological survey of North Carolina. Eventually Emmons' situation grew almost intolerable:

> I am so perfectly afloat in political feeling, that I am nearly useless. I have to be cautious in my words but I can't conceal my union feelings. Madness rules the hour & most absurdly like the nation of Israel in the days of *Rehoboam* we are a divided people. . . . It is not safe for a northern man to register his name on the books of the hotel, as a native even of N.Y. There is more tyranny in S.C. than in Austria. This state stands by the Union, though you will see much which appears to the contrary. She will go however with Virginia & other border states.[46]

Although he might yet have fled to the North before the outbreak of the Civil War, Emmons chose to stay in North Carolina, where he died in 1863 at the age of sixty-four.

Conclusions

The geological career of Ebenezer Emmons did not come to a close in the North with the damaging verdict of the Foster trial, as implied by Clarke and others.[47] Instead, a second full and useful career was launched in the South. Emmons was the first geologist to argue for the existence of a "primordial" system in America, and he continued this battle in the South. Moving beyond the confinements of the Taconic controversy, however, Emmons stirred fresh controversy with regard to Mesozoic stratigraphy. He became the first geologist to recognize the Triassic age of the Newark Group. By no means was Emmons infallible in his work, but whatever else might be said of him, he steadfastly defended his own opinions in the face of tremendous opposition.

The disputes of another time no longer bear directly on the principal geological questions of today, but it is important for us to be reminded of our potential for insensitivity to those who do not conform to popularly held scientific beliefs.

Acknowledgments

Special thanks go to Martha Riley Johnson, of the Williams College Library, for making manuscript materials in the Williamsiana Collection available to me, and for making many other valuable contributions to this study. Karl Kabelac, of the University of Rochester Library, cooperated in making accessible additional materials held in the Department of Rare Books, Manuscripts and Archives. Acknowledgments are also due to Niles Eldredge of the American Museum of Natural History and to William A. Oliver, Jr., and Clayton E. Ray of the National Museum of Natural History, Smithsonian Institution, for their insight regarding taxonomic questions and help in suggesting pertinent literature.

Notes

1. M. Hopkins, "Sketch of Dr. Emmons," *The Williams Quarterly* 11 (1864):260–69; J. Marcou, "Biographical Notice of Ebenezer Emmons," *American Geologist* 7 (1891):1–23.

2. C. J. Schneer, "Ebenezer Emmons and the Foundations of American Geology," *Isis* 60 (1969):439–50; G. P. Merrill, *The First One Hundred Years of American Geology* (New Haven: Yale University Press, 1924); and C. J. Schneer, "The Great Taconic Controversy," *Isis* 69 (1978):173–91.

3. J. M. Clarke, *James Hall of Albany: Geologist and Paleontologist, 1811–1898* (Albany, 1921).

4. Hopkins, "Sketch," p. 265.

5. E. Emmons, *Agriculture of New York*, vol. 1 (Albany: C. Van Benthuysen and Co., 1846); idem, *The Taconic System Based on Observations in New York, Massachusetts, Maine, Vermont, and Rhode Island* (Albany: Carroll and Cook, 1844).

6. Hopkins, "Sketch," pp. 261–62.

7. E. McAllister, *Amos Eaton: Scientist and Educator, 1776–1842* (Philadelphia: University of Pennsylvania Press, 1941), p. 568.

8. M. E. Johnson, "Geology in American Education: 1825–1860," *Bulletin of the Geological Society of America* 88 (1977):1192–98.

9. E. Emmons, *Geological Report of the Midland Counties of North Carolina* (New York: George P. Putnam and Co., 1856), pp. 10–12.

10. Ibid., p. 18.

11. E. Emmons, *Report of the North Carolina Geological Survey: Agriculture of the Eastern Counties* (Raleigh: H. D. Turner, 1858).

12. E. Emmons, *American Geology*, vols. 1 and 2 (Albany: Sprague and Co., 1855).

13. Ibid., p. 70.

14. Ibid., p. 38.

15. J. B. Hadley, "The Ocoee Series and Its Possible Correlatives," in *Studies of Appalachian Geology: Central and Southern*, ed. G. W. Fisher, F. J. Pettijohn, and J. C. Reed, Jr. (New York: John Wiley and Sons, 1970).

16. J. Rodgers, *The Tectonics of the Appalachians* (John Wiley and Sons, 1970).

17. Emmons, *American Geology*, 1:30–36.

18. Ebenezer Emmons to Chester Dewey, December 6, 1855, Chester Dewey Papers, University of Rochester Library.

19. H. W. Sundelius, "The Carolina Slate Belt," in *Studies of Appalachian Geology: Central and Southern*, ed. G. W. Fisher, F. J. Pettijohn, and J. C. Reed, Jr. (New York: John Wiley and Sons, 1970); W. C. Overstreet, "The Piedmont in South Carolina," in ibid.

20. E. Emmons, *Manual of Geology* (Philadelphia: Sower, Barnes and Co., 1860).

21. Rodgers, *Tectonics*, p. 165.

22. Emmons, *Taconic System*.

23. J. M. Bird and F. Rasetti, *Lower, Middle, and Upper Cambrian Faunas in the Taconic Sequence of Eastern New York: Stratigraphic and Biostratigraphic Significance*, Geological Society of America Special Paper no. 113 (1968).

24. Emmons, *American Geology*, 2:113.

25. Merrill, *First Hundred Years*, pp. 594–614.

26. N. Eldredge, personal communication, 1979.

27. Emmons, *Midland Counties*, p. xvi; and idem, "On New Fossil Corals from North Carolina," *American Journal of Science*, n.s. 22 (1856): 389–90.

28. J. St. Jean, Jr., "New Cambrian Trilobite from the Piedmont of North Carolina," *American Journal of Science* 273-A (1973):196–216.

29. P. Cloud, J. Wright, and L. Glover, III, "Traces of Animal Life from 620-Million-Year-Old Rocks in North Carolina," *American Scientist* 64 (1976):396–406.

30. W. A. Oliver, personal communication, 1979.

31. Rodgers, *Tectonics*, pp. 75–77.

32. E. Emmons, *Geology of New York*, part 2, *Comprising the Survey of the Second Geological District* (Albany: W. and A. White and J. Visscher, 1842); on Emmons' relationship to Werner, see Schneer, "Ebenezer Emmons."

33. C. D. Walcott, "The Taconic System of Emmons, and the Use of the Name Taconic in Geological Nomenclature," *American Journal of Science*, 3rd ser. 35 (1888): 229–42, 307–27, 394–401.

34. P. E. Olsen, B. Cornet, and K. S. Thomson, "Cyclic Change in Late Triassic Lacustrine Communities," *Science* 201 (1978):729–33.

35. W. C. Redfield, "On the Relations of the Fossil Fishes of the Sandstone of Connecticut and Other Atlantic States with the Liassic and Oolitic Periods," *American Journal of Science*, n.s. 22 (1856):357–63.

36. Emmons, *Midland Counties*.

37. C. Dewey, "Review of *Geological Report of the Midland Counties of North Carolina*, by E. Emmons," *American Journal of Science*, n.s. 24 (1857):427–30.

38. E. Emmons, *American Geology*, part 6. (Albany: Sprague and Co., 1857), p. 57; [R. Chambers], *Vestiges of the Natural History of Creation* (London: J. Churchill, 1844).

39. Ebenezer Emmons to Albert Hopkins, September 15, 1857, Williamsiana Collection, Williams College Library.

40. Ebenezer Emmons to Mark Hopkins, October 22, 1857, Williamsiana Collection, Williams College Library.

41. W. J. Arkell, *The Jurassic System in Great Britain* (Oxford: Oxford University Press, 1933).

42. S. J. Gould, "Eternal Metaphors of Palaeontology," in *Patterns of Evolution*, ed. A. Hallam (Amsterdam: Elsevier Scientific Publishing Company, 1977).

43. C. Lyell, *Principles of Geology*, 6th ed. (Boston: Hilliard, Gray and Co., 1842), p. 231.

44. C. Lyell, *Principles of Geology*, 11th ed. (New York: D. Appleton and Co., 1873), 1:163.

45. H. F. Osborn, "A New Mammal from the American Triassic," *Science* 8 (1886):540; idem, "The Triassic Mammals *Dromatherium* and *Microconodon*," *Proceedings of the American Philosophical Society* 24 (1887):109–11; G. G. Simpson, "Mesozoic Mammalia, part 5 (*Dromatherium* and *Microconodon*)," *American Journal of Science* 212 (1926):87–108; W. G. Kühne, "Rhaetische Triconodonten aus Glamorgan, ihre Stellung zwischen den Klassen Reptilia und Mammalia und ihre Bedeutung für die Reichart'sche Theorie," *Paläontologische Zeitschrift* 32 (1958): 197–235; and J. A. Lillegraven, Z. Kielan-Jaworowska, and W. A. Clemens, eds., *Mesozoic Mammals* (Berkeley and Los Angeles: University of California Press, 1967).

46. Ebenezer Emmons to Chester Dewey, January 3, 1861, Chester Dewey papers, University of Rochester Library.

47. Clarke, *James Hall*.

Bibliography

Manuscript Sources

Currey, Richard O. "An Essay on Iron for the Degree of Doctor of Medicine in the University of Pennsylvania." January, 1840. University of Pennsylvania Library, Philadelphia, Pennsylvania.

Eccleston, Samuel, Correspondence. Eccleston Papers, Archives of the Roman Catholic Archdiocese of Baltimore, Baltimore, Maryland.

Emmons, Ebenezer, Correspondence. Chester Dewey Papers, University of Rochester Library, Rochester, New York.

Emmons, Ebenezer, Correspondence. Williams College Library, Williamstown, Massachusetts.

Gibbs Family Papers. Newport Historical Society, Newport, Rhode Island.

Lagsdorf, Alexander S. "The Story of Washington University, 1853–1953." Manuscript Collection, Missouri Historical Society, St. Louis, Missouri.

Lieber, Oscar, Correspondence. Francis Lieber Papers, South Caroliniana Library, University of South Carolina, Columbia, South Carolina.

Litton, Abram, Correspondence and Papers. Lane Collection, Missouri Historical Society, St. Louis, Missouri.

Medical College of Ohio, Records. University of Cincinnati Library, Cincinnati, Ohio.

New Harmony Community, Petty Cash Journal, Volume 2. The Workingman's Institute, New Harmony, Indiana.

Nicollet, Joseph N., Correspondence. Ford Collection, New York City Public Library, New York, New York.

Nicollet, Joseph N., Correspondence. Greer Papers, Historical Society of Pennsylvania, Philadelphia, Pennsylvania.

Nicollet, Joseph N., Correspondence. Nicollet Papers, Minnesota Historical Society, St. Paul, Minnesota.

Nicollet, Joseph N., Correspondence and Personal Papers. Library of Congress, Washington, D.C.

Posey County, Indiana, Record Book C, February Term, 1827. Posey County Court House, Mount Vernon, Indiana.

Ruffin, Edmund. "Incidents of My Life, 1821–1851." Virginia Historical Society, Richmond, Virginia.

Ruffin, Edmund. "Private Diary of Ed. Ruffin, Agricultural Surveyor, in S.C.[a], 1843." Virginia Historical Society, Richmond, Virginia.

Ruffin, Edmund, Papers. Southern Historical Collection, University of North Carolina Library, Chapel Hill, North Carolina.

Schoepflin, Gary Lee. "Denison Olmsted (1790–1859), Scientist, Teacher, Christian: A Biographical Study of the Connection of Science with Religion." Ph.D. Dissertation, Oregon State University, 1977.

Troost, Gerard. "Catalogue of a Collection of Minerals Formed by G. Troost. . . ." Vols. 1 and 2, unpaginated, about 1400 pages. Kentucky Room, Free Public Library, Louisville, Kentucky.

Troost, Gerard, Personal Papers. Tennessee State Library and Archives, Nashville, Tennessee.

University of South Carolina, Minutes of Meetings of the Board of Trustees, vol. 2, 1813–37. South Caroliniana Library, University of South Carolina, Columbia, South Carolina.

Virginia Geological Survey Papers. Virginia State Library, Richmond, Virginia. Especially the correspondence of William Aikin, George Boyd, Caleb Briggs, J. Brown, Charles B. Hayden, Henry Darwin Rogers, James B. Rogers, Robert E. Rogers, William Barton Rogers, and Israel Slade.

Rogers, William Barton, Papers. Institute Archives and Special Collections, Massachusetts Institute of Technology Library, Cambridge, Massachusetts.

Published Sources

Anonymous Publications

Reports and Resolutions of the General Assembly of South-Carolina, Passed at Its Regular Session of 1842. Columbia: A. H. Pemberton, 1843.

"Professor Litton." *Guardian* 3 (1843):40.

"Dr. Troost's Report." *Guardian* 4 (1844):21.

Reports and Resolutions of the General Assembly of the State of South Carolina, Passed at the Annual Session of 1848. Columbia: I. C. Morgan, 1848.

Catalogue of the Officers and Graduates of the University of Nashville. Nashville: A. Nelson, 1850.

"Southern Agriculturist." *Southern Lady's Companion* 4 (1851):262.

"Edwin [*sic*] Ruffin, of Virginia, Agriculturist, Embracing a View of Agricultural Progress in Virginia for the Last Thirty Years." *DeBow's Review* 11 (1851):431–36.

"Geological Survey." *South Carolina Legislative Times*. 33rd Legislature, 2nd Session. December 11, 1855, pp. 171–73.

"Prof. Tuomey." *American Journal of Science* 73 (1857):448.

Review of Oscar M. Lieber's Third Annual Report. *American Journal of Science*, 2nd ser. 28 (1859):148.

Journal of the House of Representatives of the Commonwealth of Kentucky. Regular Session of December 4, 1871. Frankfort: Kentucky Yeoman, 1871.

Journal of the House of Representatives of the Commonwealth of Kentucky. Session of December 1, 1873. Frankfort: Kentucky Yeoman, 1873.

The History of Jackson County, Missouri, Containing a History of the County, Its Cities, Towns, Etc. Kansas City, Missouri: Union Historical Co., 1881.

Annual Report of the Superintendent, United States Military Academy. Washington, 1896.

Signed Publications

Adams, Frank D. "Earliest Use of the Term Geology." *Bulletin of the Geological Society of America* 43 (1932):121.

———. "Further Notes on the Use of the Term Geology." *Bulletin of the Geological Society of America* 44 (1933):831.

Aldrich, Mark. "Earning of American Civil Engineers, 1820–1859." *Journal of Economic History* 31 (1971):407–19.

Aldrich, Michele L. "American State Geological Surveys, 1820–1845." In *Two Hundred Years of American Geology: Proceedings of the New Hampshire Bicentennial Conference on the History of Geology*. Edited by Cecil J. Schneer. Hanover, N.H.: University Press of New England, 1979.

Arkell, W. J. *The Jurassic System in Great Britain*. Oxford: Oxford University Press, 1933.

Baren, J. van. *Nieuw Nederlandsch biographisch woordenboek*. Vol. 5. Leiden, 1921.

Bartram, John. "Diary of a Journey through the Carolinas, Georgia and Florida from July 1, 1765, to April 10, 1766. . . ." Annotated by Francis Harper. *Transactions of the American Philosophical Society*, n.s. 33, no. 1 (1942):1–120.

Bartram, William. *Travels through North and South Carolina, Georgia, East and West Florida*. . . . Philadelphia: James and Johnson, 1791.

———. "Travels in Georgia and Florida, 1773–74: A Report to Dr. John Fothergill." Annotated by Francis Harper. *Transactions of the American Philosophical Society*, n.s. 33, no. 2 (1943):121–242.

Baumgartel, Hans. "Alexander von Humboldt." In *Toward a History of Geology*. Edited by Cecil J. Schneer. Cambridge, Mass.: The MIT Press, 1969.

Bestor, Arthur E., Jr. "Education and Reform at New Harmony: Correspondence of William Maclure and Marie Duclos Fretageot, 1820–1833." *Indiana Historical Collection* 15 (1948):285–417.

Bidwell, Percy W., and John I. Falconer. *History of Agriculture in the*

Northern United States, 1620–1860. Clifton, N.J.: Augustus M. Kelly Publishers, 1973.

Bird, J. M., and F. Rasetti. *Lower, Middle and Upper Cambrian Faunas in the Taconic Sequence of Eastern New York: Stratigraphic and Biostratigraphic Significance*. Geological Society of America Special Paper no. 113, 1968.

Bray, Martha C. "Joseph Nicolas Nicollet, Geologist." *Proceedings of the American Philosophical Society* 114 (1970):37–59.

Broadhead, G. C. "Abram Litton, M.D." *Transactions of the St. Louis Academy of Science* 12 (1902):xxiv–xxvii.

Brown, L. A. *Early Maps of the Ohio Valley . . . from 1673 to 1783*. Pittsburgh: University of Pittsburgh Press, 1959.

Brown, Paul. *Twelve Months in New Harmony, Representing a Faithful Account of the Principle Occurrences Which Have Taken Place within That Period, Interspersed with Remarks*. Cincinnati: William Hill Woodward, 1827.

Burgess, George. *A Journal of a Surveying Trip into Western Pennsylvania, under Andrew Ellicott in the year 1795*. . . . Edited by John Cumming. Mt. Pleasant, Mich.: John Cumming, 1965.

Chambers, R. *Vestiges of the Natural History of Creation*. London: J. Churchill, 1844.

Clark, T. D. *Travels in the Old South: A Bibliography*. Vol. 2. Norman: University of Oklahoma Press, 1956.

Clarke, John M. *James Hall of Albany, Geologist and Palaeontologist 1811–1898*. Albany, 1921.

Clemson, Thomas G. "Fertilizers." In *Report of the Commissioner of Patents for the Year 1859, Agriculture*. Washington, D.C.: George W. Bowman, 1860.

Cloud, P., J. Wright, and L. Glover, III. "Traces of Animal Life from 620-Million-Year-Old Rocks in North Carolina." *American Scientist* 64 (1976):396–406.

Corgan, James X. "Tennessee's Early Technical and Scientific Journals, 1825–1861." *Journal of the Tennessee Academy of Science* 52 (1977):23–26.

———. "Offbeat Pioneers." *Journal of the Tennessee Medical Association* 70 (1977):873–76.

———. "Some Firsts? in the Colleges of Tennessee." *Journal of the Tennessee Academy of Science* 55 (1980):86–91.

———. "A Preliminary Catalog of the Scientific and Technical Societies of Antebellum East Tennessee." *East Tennessee Historical Society's Publications* 48 (1980 for 1976):21–28.

Crocker, William. *The History of Agricultural Gypsum*. Chicago: The Gypsum Industries Association, 1922.

Currey, Richard O. "Geology of Benton County, Ala." *Southern Journal of Medical and Physical Sciences* 2 (1854):195–202.

———. *A Sketch of the Geology of Tennessee*. Knoxville: Haskell, 1857.

———. *A Geological Visit to the Virginia Copper Region*. Knoxville: Beckett, Haws and Co., 1859.

———. "The Paper on the Copper and Other Resources of the Floyd-Carroll-Grayson Plateau." *The Virginias* 1 (1880): 60, 62–64, 69–71, 74–77, 80–81, 95.

Cutter, W. P. "A Pioneer in Agricultural Science." In *Yearbook of the United States Department of Agriculture*, 1895. Washington, D.C.: U.S. Government Printing Office, 1896.

Dewey, C. "Review of 'Geological Report of the Midland Counties of North Carolina' by E. Emmons." *American Journal of Science*, n.s. 24 (1857): 427–30.

Dunbar, William. "Description of the River Mississippi and Its Delta." *Transactions of the American Philosophical Society* 6, no. 1 (1809):165–87.

Eaton, Clement. *The Mind of the Old South*. Rev. ed. Baton Rouge: Louisiana State University Press, 1964.

Ellicott, Andrew. "Description of the Falls of Niagara." *Universal Asylum and Columbian Magazine* 4 (1790):331–32, plate.

———. "Miscellaneous Observations relative to the Western Part of Pennsylvania, Particularly Those in the Neighborhood of Lake Erie." *Transactions of the American Philosophical Society* 4 (1799):224–29.

———. *The Journal of . . . Late Commissioner on Behalf of the United States during Part of the Year 1796, the Years 1797, 1798, 1799, and Part of the Year 1800: For Determining the Boundary between the United States and the Possessions of His Catholic Majesty in America, Containing Occasional Remarks on the Situation, Soil, Rivers, Natural Productions, and Diseases of the Different Countries of the Ohio, Mississippi and the Gulf of Mexico*. Philadelphia: Budd and Bartram for Thomas Dobson, 1803.

Elliott, Clark. *Biographical Dictionary of American Science: The Seventeenth through the Nineteenth Centuries*. Westport, Conn.: Greenwood, 1979.

Emmons, Ebenezer. *Geology of New York*. Part 2. *Comprising the Survey of the Second Geological District*. Albany: W. and A. White and J. Visscher, 1842.

———. *The Taconic System*. Albany: Carroll and Cook, 1844.

———. *American Geology*. Vols. 1 and 2. Albany: Sprague and Co., 1855.

———. *Geological Report of the Midland Counties of North Carolina*. New York: C. P. Putnam and Co., 1856.

———. "On New Fossil Corals from North Carolina." *American Journal of Science*, n.s. 22 (1856):389–90.

———. *American Geology*, Part 4. Albany: Sprague and Co., 1857.

———. *Manual of Geology*. Philadelphia: Sower, Barnes and Co., 1860.

Evans, Lewis. *An Analysis of the General Map of the Middle British Colonies in America*. Philadelphia: B. Franklin and D. Hall, 1755.

Eve, Paul F. "Brief Biographical Sketches of the Professional Dead of Middle Tennessee." *Transactions of the Medical Society of the State of Tennessee* (1876), pp. 78–90.

Faust, Drew Gilpin. *A Sacred Circle: The Dilemma of the Intellectual in the Old South, 1840–1860*. Baltimore: Johns Hopkins University Press, 1977.

Featherstonhaugh, George W. *Excursion through the Slave States, from Washington on the Potomac to the Frontiers of Mexico; with Sketches of Popular Manners and Geological Notices*. New York: Harper and Brothers, 1844.

———. *The Monthly American Journal of Geology and Natural Science*, Vol. 1 *1831–1832*. Reprinted with an introduction by G. W. White. New York: Hafner Press, 1969.

Filson, John. *The Discovery, Settlement and Present State of Kentucky*. Wilmington, Del.: James Adams, 1784.

Flint, Timothy. *A Condensed Geography and History of the Western States of the Mississippi Valley*. Cincinnati: E. H. Flint, 1828.

Franklin, Benjamin. "A Letter From Dr. Benjamin Franklin to Mr. Alphonsus le Roy . . . Containing Sundry Maritime Observations." *Transactions of the American Philosophical Society* 2 (1786):294–329.

Gardener, Julia. *Mollusca from the Miocene and Lower Pliocene of Virginia and North Carolina*. United States Geological Survey Professional Paper no. 199, 1943.

Gerstner, Patsy. "A Dynamic Theory of Mountain Building: Henry Darwin Rogers, 1842." *Isis* 66 (1975):26–37.

———. "Henry Darwin Rogers and William Barton Rogers on the Nomenclature of the American Paleozoic Rocks." In *Two Hundred Years of American Geology: Proceedings of the New Hampshire Bicentennial Conference on the History of Geology*. Edited by Cecil J. Schneer. Hanover, N.H.: University Press of New England, 1979.

Glenn, L. C. "Gerard Troost." *American Geologist* 35 (1905):72–94.

———. "The Growth of Our Knowledge of Tennessee Geology." *Tennessee Geological Survey, Resources of Tennessee* 2 (1912):167–219.

Goldstein, Max. *One Hundred Years of Medicine and Surgery in Missouri*. St. Louis: St. Louis Star, 1900.

Goodrich, W. H. *The History of the Medical Department of the University of Georgia*. Augusta, n.d.

Goodwin, E.J. *A History of Medicine in Missouri*. St. Louis: W. L. Smith, 1905.

Gould, Stephen Jay. "Eternal Metaphors of Palaeontology." In *Patterns of Evolution*. Edited by A. Hallam. Amsterdam: Elsevier Scientific Publishing Co., 1977.

Graham, Gerald S. "The Gypsum Trade of the Maritime Provinces." *Agricultural History* 12 (1938):209–23.

Gray, Lewis C. *History of Agriculture in the Southern United States to 1860*. Clifton, N.J.: Augustus M. Kelly, 1973.

Hadley, J. B. "The Ocoee Series and Its Possible Correlatives." In *Studies of Appalachian Geology: Central and Southern*. Edited by G. W. Fisher, F. J. Pettijohn, and J. C. Reed, Jr. New York: John Wiley and Sons, 1970.

Hamy, E. T. *The Travels of the Naturalist Charles A. Lesueur in North America*. Translated by Milton Haber and edited by H. F. Raup. Kent, Oh.: Kent State University Press, 1975.

Harland, W. B. *The Phanerozoic Time-Scale*. London: The Geological Society of London, 1964.

Hazen, Robert M. *Bibliography of American Published Geology: 1669–1850*. Geological Society of America Microform Publication no. 4, 1976.

———, Margaret Hindle Hazen. *American Geological Literature, 1669–1850*. Stroudsburg, Penn.: Dowden, Hutchinson, and Ross, 1980.

Hendrickson, Walter B. *David Dale Owen, Pioneer Geologist of the Middle West*. Indianapolis: Indiana Historical Bureau, 1943.

Herndon, G. M. "Agricultural Reform in Antebellum Virginia: William Galt, Jr., A Case Study." *Agricultural History* 52 (1978):394–406.

Holmes, Francis S. *Phosphate Rocks of South Carolina and the Great Carolina Marl Bed*. Charleston: Holmes' Book House, 1870.

Holmes, J. A. "Mineralogical, Geological and Agricultural Surveys of South Carolina." *Journal of the Elisha Mitchell Scientific Society* 7 (1890):89–117.

Hopkins, M. "Sketch of Dr. Emmons." *The Williams Quarterly* 11 (1864): 260–69.

Hough, Franklin B., ed. *American Biographical Notes*. Albany: Jack Munsell, 1875.

Hutchins, Thomas. *An Historical Narrative and Topographical Description of Louisianna and West Florida, Comprehending the River Mississippi with its Principal Branches and Settlements*. Philadelphia: for the author, 1784.

———. *A Topographical Description of Virginia, Pennsylvania, Maryland, and North Carolina, Comprehending the Rivers Ohio, Kenhawa, Sioto, Cherokee, Wabash, Illinois, Mississippi. . . .* London: J. Almon, 1778.

Hyde, J. E. *The Mississippian Formations of Central and Southern Ohio*. Edited by M. R. Marple. Ohio Geological Survey Bulletin no. 51, 1953.

Ihde, Aaron J. "Edmund Ruffin, Soil Chemist of the Old South." *Journal of Chemical Education* 29 (1952):407–14.

Imlay, Gilbert. *Topographical Description of the Western Territory of North America*. 3rd ed. London: J. Debrett, 1797.

Jacob, K. D. "Predecessors of Superphosphate." *Superphosphate: Its History, Chemistry, and Manufacture*. Washington: U.S. Government Printing Office, 1964.

Jillson, Willard Rouse. *Geological Research in Kentucky*. Kentucky Geological Survey Bulletin, 6th ser., no. 15, 1923.

———. "The Advance of Geology in Kentucky During One Hundred and Thirty Four Years (1798–1918)." *Transactions of the Kentucky Academy of Science* 25 (1965):48–57.

Johnson, Markes E. "Early Explorations of the Silurian System in Iowa." *Proceedings of the Iowa Academy of Science* 84 (1977):150–56.

———. "Geology in American Education: 1825–1860." *Bulletin of the Geological Society of America* 88 (1977):1192–98.

Johnson, Thomas Cary, Jr. *Scientific Interests in the Old South*. New York: Appleton-Century, 1936.

Kennedy, J. P. *Memoirs of William Wirt*. Vols. 1 and 2. Philadelphia: Lea and Blanchard, 1849.

Kühne, W. G. "Rhaetische Triconodonten aus Glamorgan, ihre Stellung zwischen den Klassen Reptilia und Mammalia und ihre Bedeutung für die Reichart'sche Theorie." *Paläontologische Zietschrift* 32 (1958):197–235.

LaBorde, Maximilian. *History of South Carolina College, from Its Incorporation December 19, 1801, to Nov. 25, 1857, Including Sketches of Its Presidents and Professors*. Columbia: Peter B. Glass, 1859.

Lieber, Oscar M. *Report on the Survey of South Carolina: Being the First Annual Report to the General Assembly of South Carolina, Embracing the Progress of the Survey during the Year 1857; with Plates and Maps*. 2nd ed. Columbia: R. W. Gibbes, 1858.

———. *Report on the Survey of South Carolina: Being the Second Annual Report to the General Assembly of South Carolina, Embracing the Progresses of the Survey during the Year 1857; with Plates and Maps*. Columbia: R. W. Gibbes, 1858.

———. [Metallicus]. "The Survey of South Carolina: No. 1." *Charleston Mercury*, October 24, 1855.

———. [Metallicus]. "The Proposed Survey of the State: No. 2." *Charleston Mercury*, October 25, 1855.

———. [Metallicus]. "The Technico-Geological Survey of South Carolina." *Charleston Mercury*, November 16, 1855.

Lillegraven, J. A., Z. Kielan-Jaworowska, and W. A. Clemens. *Mesozoic Mammals*. Berkeley and Los Angeles: Univ. California Press, 1979.

Litton, A. "Agricultural Education." *Agriculturist* 2 (1841):139.

———. "Chemistry." *Agriculturist* 5 (1844): 33–35, 49–50, 65–67.

Lonsdale, John T. *Geology of the Gold-Pyrite Belt of the Northeastern Piedmont Virginia*. Virginia Geological Survey Bulletin no. 30, 1927.

Lyell, Charles. *Principles of Geology*. 6th ed. Boston: Hilliard, Gray and Co., 1842.

———. "On the Tertiary Formations and Their Connexion with the Chalk in Virginia and Other Parts of the United States." *Proceedings of the Geological Society of London* 3 (1842):735–42.

———. "On the Geological Position of the *Mastodon giganteum* and Associated Fossil Remains at Bigbone Lick, Kentucky, and Other Localities in the United States and Canada." *Proceedings of the Geological Society of London* 4 (1843):33–39. Reprinted in *American Journal of Science* 46 (1844):320–23.

———. *Travels in North America in the Years 1841–2; with Geological Observations on the United States, Canada and Nova Scotia*. 2 vols. New York: Wiley and Putnam, 1845.

———. "Miocene Formations of Virginia and North Carolina, etc., with Appendix, on Fossil Corals, by Mr. Lonsdale." *Proceedings of the Geological Society of London* 4 (1845):547–63.

———. "On the White Limestone of South Carolina and Georgia, and the Eocene Strata of Other Parts of the U.S., with Appendix, on the Corals, by Mr. Lonsdale." *Proceedings of the Geological Society of London* 4 (1845): 563–76.

———. "Notes on the Cretaceous Strata of New Jersey and Other Parts of the United States Bordering the Atlantic." *Quarterly Journal of the Geological Society of London* 1 (1845):55–60.

———. "Observations on the White Limestone and Other Eocene or Older Tertiary Formations of Virginia, Carolina, and Georgia." *Quarterly Journal of the Geological Society of London* 1 (1845):429–42.

———. "Claiborne, Ala., Dated Feb. 4, 1846, on the *Zeuglodon* Fossils and Eocene Limestone of Washington and Clarke Counties, Ala." *American Journal of Science*, 2nd ser. 1 (1846):313–15.

———. "Mobile, Ala., Feb. 19, 1846. Coal Field of Tuscaloosa, Alabama." *American Journal of Science*, 2nd ser. 1 (1846):371–76.

———. "Observations on the Fossil Plants, of the Coal Field of Tuscaloosa,

Alabama; with a Description of Some Species by C. T. F. Bunbury." *American Journal of Science,* 2nd ser. 2 (1846):228–33.

———. "On the Newer Deposits of the Southern States of North America." *Quarterly Journal of the Geological Society of London* 2 (1846):405–10.

———. "On the Delta and Alluvial Deposits of the Mississippi, and Other Points in the Geology of North America, Observed in the Years 1845, 1846. . . ." *American Journal of Science,* 2nd ser. 3 (1847):34–39.

———. "On the Alleged Coexistence of Man and the *Megatherium.* . . ." *American Journal of Science,* 2nd ser. 3 (1847):267–69. Also published in the London *Times,* December 8, 1846.

———. ". . . On the Structure and Probable Age of the Coal Field of the James River, near Richmond, Virginia." *American Journal of Science,* 2nd ser. 4 (1847):113–15.

———. "On the Relative Age and Position of the So-Called Nummulite Limestone of Alabama." *American Journal of Science,* 2nd ser. 4 (1847): 186–91.

———. *A Second Visit to the United States of North America.* Vols. 1 and 2. New York: Harper, 1849. 2nd ed. rev. and corrected. London: John Murray, 1850.

———. "On the Nummulite Limestone of Alabama (Proceedings Geol. Soc., Jan. 9, 1847)." *American Journal of Science,* 2nd. ser. 7 (1849):114.

———. *Principles of Geology.* 11th ed. New York: D. Appleton and Co., 1873.

Lyell, Katherine M. *Life, Letters, and Journals of Sir Charles Lyell, Bart.* 2 vols. London: J. Murray, 1881.

McDermott, John Francis. "Dr. Koch's Wonderful Fossils." *Missouri Historical Society Bulletin* 4 (1948):233–56.

McDonald, Angus. *Early American Soil Conservationists.* U.S. Department of Agriculture Miscellaneous Publication no. 449, 1941.

[McElrath, Thomas]. Review of Oscar M. Lieber's First Annual Report. *Mining and Statistic Magazine* 10 (1858):175.

Maclure, William. "Observations on the Geology of the United States, Explanatory of a Geological Map." *Transactions of the American Philosophical Society* 6 (1809):411–28. Rev. ed. Philadelphia: Abram Small, 1817.

Marcou, J. "Biographical Notice of Ebenezer Emmons." *American Geologist* 7 (1891):1–23.

Martin, David Strobel. "The Mineralogical and Geological Cabinets of South Carolina College." *South Carolina College Bulletin* 4 (1906):11–22.

Mathews, C. V. C. *Andrew Ellicott, His Life and Letters.* New York: Grafton Press, 1908.

Mehring, A. L. "Fertilizers." In *Handbook of Food and Agriculture*. Edited by Fred C. Blanck. New York: Reingold Publishing Co., 1955.

———, J. Richard Adams, and K. D. Jacob. *Statistics on Fertilizers and Lining Materials in the United States*. Statistical Bulletin no. 191. U.S. Department of Agriculture, 1957.

Meisel, Max. *A Bibliography of American Natural History: The Pioneer Century, 1769–1865*. Vols. 1–3. New York: Premier Publishing Co., 1924, 1926, 1927. Rpt. New York: Hafner Publishing Co., 1967.

Merrill, George Perkins. "Contributions to a History of American State Geological and Natural History Surveys." *U.S. National Museum Bulletin* 109 (1920):1–549.

———. *The First One Hundred Years of American Geology*. New Haven: Yale University Press, 1924. Rpt. New York: Hafner Publishing Co., 1964.

Millbrooke, Anne, "The Geological Society of Pennsylvania, 1832–1836." *Pennsylvania Geology* 7 (1976):7–11 and 8 (1977):12–16.

Mitchell, Samuel L. "Observations on the Geology of North America. . . ." In *Essay on the Theory of the Earth*. Edited by G. Cuvier and translated by J. Jameson. New York: Kirk and Mercein, 1818.

Morse, Zedidiah. *The American Geography*. Elizabethtown, N.J.: Shephard Kollick, 1789.

Nelson, Katherine G. "Environment for Discovery: The Owen Survey of Wisconsin." *Wisconsin Academy of Science, Arts, and Letters* 64 (1976): 173–79.

Nickles, John Milton. *Geologic Literature on North America: 1785–1918*. U.S. Geological Survey Bulletin no. 746, 1923.

Nicollet, J. N. *Report Intended to Illustrate a Map of the Hydrographic Basin of the Upper Mississippi River*. Washington, 26th Congress, 2nd Session, Senate Document 237, serial 380, 1843.

Olsen, P. E., P. Cornet, and K. S. Thomson. "Cyclic Change in Late Triassic Lacustrine Communities." *Science* 201 (1978):729–33.

Osborn, H. F. "A New Mammal from the American Triassic." *Science* 8 (1886):540.

———. "The Triassic Mammals *Dromatherium* and *Microconodon*." *Proceedings of the American Philosophical Society* 24 (1887):109–11.

Overstreet, W. C. "The Piedmont in South Carolina." *Studies of Appalachian Geology: Central and Southern*. Edited by G. W. Fisher, F. J. Pettijohn, and J. C. Reed. New York: John Wiley and Sons, 1970.

Owen, David Dale, et al. *Report of the Geological Survey in Kentucky*. Frankfort: J. B. Major, 1861.

Platt, Samuel J., and Mary L. Ogden. *Medical Men and Institutions of Knox*

County, Tennessee, 1789–1957. Knoxville: S. B. Newman Printing Co., 1969.

Porter, Roy. "Gentlemen and Geology: The Emergence of a Scientific Career, 1660–1920." *Historical Journal* 21 (1978):809–36.

Rabbitt, Mary C. *Minerals, Land, and Geology for the Common Defense and General Welfare*. Vol. 1, before 1879. Washington, D.C.: U.S. Geological Survey, 1979.

Rasmussen, Wayne D. *Agriculture in the United States: A Documentary History*. New York: Random House, 1975.

Redfield, W. C. "On the Relations of the Fossil Fishes of the Sandstone of Connecticut and other Atlantic States to the Liassic and Oolitic Periods." *American Journal of Science*, n.s. 22 (1856):357–63.

Reingold, Nathan. *Science in Nineteenth-Century America: A Documentary History*. American Century Series. New York: Hill and Wang, 1964.

Richardson, P. L. "Benjamin Franklin and Timothy Folger's First Printed Chart of the Gulf Stream." *Science* 207 (1980):643–45.

Roberts, Joseph K. "William Barton Rogers and His Contribution to the Geology of Virginia." *Virginia Geological Survey Bulletin* 46 (1936):25–28.

Robinson, L. C. "The Kentucky Geological Survey." *Register of the Kentucky State Historical Society* 25 (1927):86–93.

Rodgers, J. *The Tectonics of the Appalachians*. New York: John Wiley and Sons, 1970.

Rogers, Emma Savage. *Life and Letters of William Barton Rogers*. Vols. 1 and 2. Boston: Houghton Mifflin, 1896.

Rogers, William Barton. "Report of the Geological Reconnoissance of the State of Virginia." *Virginia State Legislature, House of Delegates Documents* 24 (1836):1–52. Republished in book form, Philadelphia: Desilver, Thomas, and Co., 1836 (1837).

———. "On the Connection of Thermal Springs in Virginia with Anticlinal Axes and Faults." *Transactions of the American Association of Geologists and Naturalists*, 1843, pp. 323–47.

———. Tuomey obituary. *Proceedings of the Boston Society of Natural History* 6 (1857):184–86.

———. *A Reprint of the Annual Reports and Other Papers on the Geology of the Virginias*. New York: Appleton, 1884.

———. Henry Darwin Rogers. "On the Physical Structure of the Appalachian Chain as Exemplifying the Laws Which have Regulated the Elevation of Great Mountain Chains Generally." *Transactions of the American Association of Geologists and Naturalists*, 1843, pp. 475–531.

Rooker, H. G. "A Sketch of the Life and Works of Dr. Gerard Troost." *Tennessee Historical Magazine*, 2nd ser. 3 (1933):3–19.

Ross, Sydney. "Scientist: The Story of A Word." *Annals of Science* 18 (1962):65–85.

Rossiter, Margaret W. *The Emergence of Agricultural Science: Justus Liebeg and the Americans, 1840–1880*. New Haven: Yale University Press, 1975.

Royer, A. *History and Description of the Royal Museum of Natural History*. Paris: Cellet, 1823.

Rozet, M. *Cours Elementaire de geognosie*. Paris, 1830.

Rudwick, Martin J. J. "The Emergence of a Visual Language for the Geological Sciences, 1760–1840." *History of Science* 14 (1976):149–95.

Ruffin, Edmund. *An Essay on Calcareous Manures*. Cambridge: Harvard University Press, Belknap Press, 1961.

Schneer, Cecil J. "Ebenezer Emmons and the Foundations of American Geology." *Isis* 60 (1969):439–50.

———. "The Great Taconic Controversy." *Isis* 69 (1978):173–91.

Schoepf, J. D. *Beyträge zur mineralogischen Kenntniss des ostlichen Theils von Nordamerika und seiner Gebürge*. Erlangen: J. J. Palm, 1787. Facsimile and annotated translation by E. M. Spieker, *Geology of Eastern North America*. New York: Hafner Publishing Company, 1972.

Shaler, Nathaniel Southgate. "Report on the Unfinished Work of the Survey of the Commonwealth, under the Direction of Dr. David Dale Owen." *Reports on the Progress of the Kentucky Geological Survey* 3, no. 8 (1872):417–20.

———. Preface to *Reports of Progress of the Kentucky Geology Survey*. 2nd ser., vol. 1. 1876.

Shrock, Robert Rakes. *Geology at M.I.T., 1865–1965: A History of the First One Hundred Years of Geology at the Massachusetts Institute of Technology*. Cambridge: MIT Press, 1977.

Silliman, Benjamin. "Remarks on Some of the Gold Mines and on Parts of the Gold Region of Virginia. Founded on Personal Observations Made in the Months of August and September, 1836." *American Journal of Science*, 1st ser. 32 (1837):98–130.

Simpson, G. G. "Mesozoic Mammalia. Part V (*Dromatherium* and *Microconodon*)." *American Journal of Science* 212 (1926):87–108.

Smith, Eugene A. "Sketch of the Life of Michael Tuomey." *American Geologist* 20 (1897):204–12.

Snow, C. P. *The Two Cultures* and *The Scientific Revolution*. Cambridge: At the University Press, 1959.

St. Jean, J., Jr. "New Cambrian Trilobite from the Piedmont of North Carolina," *American Journal of Science* 273-A (1973):196–216.

Stork, William. *An Account of East Florida; Third Edition, with a Journal Kept by John Bartram of Philadelphia . . . Botanist to His Majesty for the Floridas, Illustrated with a Map of East Florida*. London, 1769.

Sundelius, H. E. "The Carolina Slate Belt." In *Studies of Appalachian Geology: Central and Southern*. Edited by G. W. Fisher, F. J. Pettijohn, and J. C. Reed, Jr. New York: John Willey and Sons, 1970.

Swem, Earl G. "An Analysis of Ruffin's Farmers' Register, with a Bibliography of Edmund Ruffin." *Bulletin of the Virginia State Library* 11 (1918): 40–144.

Sydnor, Charles S. "State Geological Surveys in the Old South." In *American Studies in Honor of William Kenneth Boyd*. Edited by David Kelly Jackson. Durham, N.C.: Duke University Press, 1940.

Troost, Gerard, trans. *Natur-tafereelen, met weterschappelijke ophelderingen*, by Alexander von Humboldt. The Hague, 1808.

Troost, Lewis. *Report on the Preliminary Surveys of the Mobile and New Orleans Railroad*. Mobile, 1854.

Tuomey, M. *Report on the Geological and Agricultural Survey of the State of South Carolina*. Columbia: A. J. Johnston, 1844.

———. *Report on the Geology of South Carolina*. Columbia: A. J. Johnston, 1848.

Vanuxem, Lardner. Report on a Mineralogical and Geological Examination of the State of South Carolina. In *Statistics of South Carolina, Including a View of Its Natural, Civil, and Military History, General and Particular*. Compiled by Robert Mills. Charleston: Hurlbut and Lloyd, 1826.

———, and S. G. Morton. "Geological Observations on the Secondary, Tertiary, and Alluvial Formations of the Atlantic Coast of the United States of America." *Journal of the Philadelphia Academy of Sciences* 6 (1829), pt. 1, pp. 59–71.

Volney, C. F. *Tableau du climat et du sol des États-Unis d'Amérique, suivi d'éclaircissemens sur la Floride, sur la colonie française au Scioto, sur quelques colonies canadiennes et sur les sauvages*. Paris: Courcier, 1803. Translated into English by C. B. Brown as *A View of the Soil and Climate of the United States of America*. . . . Philadelphia: J. Conrad and Co., 1804. Reprinted with an introduction by G. W. White. New York: Hafner Publishing Company, 1968.

Waggaman, William H. "A Report on the Phosphate Fields of South Carolina." *Commercial Fertilizers* 7 (1913):33.

Walcott, Charles D. "The Taconic System of Emmons, and the Use of the

Name Taconic in Geological Nomenclature." *American Journal of Science*, 3rd ser. 35 (1888): 229–42, 307–27, 394–401.

Weaver, Fred B. "Medical Care in a Confederate Prison." *North Carolina Medical Journal* 25 (1964):206–9.

White, George W. "Birth of American Hydrogeology." In *History of Hydrology in the United States*. U.S. Geological Survey Professional Paper, in press.

Wilson, Charles W., Jr. *Annotated Bibliography of the Geology of Tennessee*. Tennessee Division of Geology Bulletin no. 59, 1953.

Wilson, Leonard G. *Charles Lyell, the Years to 1841: The Revolution in Geology*. New Haven: Yale University Press, 1972.

Woodward, Herbert. "Reappraisal of Appalachian Geology." *Bulletin of the American Association of Petroleum Geologists* 45 (1961):1625–33.

Zittle, Karl A. von. *History of Geology and Palaeontology to the End of the Nineteenth Century*. London: Walter Scott, 1901.

Contributors

DR. MICHELE L. ALDRICH directs the women's project at the American Association for the Advancement of Science in Washington, D.C., and is a research associate of the California Academy of Science. She has published several articles on the history of American geology.

DR. DANIEL D. ARDEN is chairman of the Department of Geology and Physics at Georgia Southwestern College, Americus, Georgia. His field of specialization is the geology of the Gulf of Mexico–Caribbean region and the southeastern U.S. coastal plain.

MARTHA COLEMAN BRAY was in charge of the Minneapolis History Collection in the Minneapolis Public Library when she first read Nicollet's *Report* and became fascinated with the little-known author. Delving deeper into Nicollet's career lead to the present, and other, studies.

DR. JAMES X. CORGAN is chairman of the Department of Geology, Austin Peay State University, Clarksville, Tennessee. He is a frequent author of articles on the history of science in the South.

DR. MARKES E. JOHNSON teaches geology at Williams College, Williamstown, Massachusetts. He is a regular contributor to the literature on the history of geology.

DR. ALAN E. LEVITON is chairman of the Department of Herpetology of the California Academy of Science in San Francisco. As a historian of science, he is principally interested in western North America.

ANNE MILLBROOKE is an assistant historian and archivist with United Technologies, in East Hartford, Connecticut. She is also a doctoral candidate in the Department of History and Sociology of Science, University of Pennsylvania, Philadelphia. Her dissertation is on several of the state geological surveys of the nineteenth century.

RICHARD C. SHERIDAN is a research chemist at the TVA National Fertilizer Development Center, Muscle Shoals, Alabama. He is the author of several papers pertaining to the history of chemistry and has served as president of the Tennessee Valley Historical Society.

GEORGE W. WHITE is Research Professor of Geology Emeritus at the University of Illinois, Champaign-Urbana. His own professional work in geology and that of his many students has been in glacial and economic geology. He has published over twenty-five papers in the history of geology, mainly in the history of early American geology. He is past vice-president of the International Committee for History of Geology and past chairman of the U.S. National Committee for History of Geology.

IVAN L. ZABILKA is assistant registrar at the University of Missouri at Columbia. He has two advanced degrees in theology and two in the history of science. His primary focus of study is the interaction of theologians and scientists in the nineteenth century.

Index

Community and county names are alphabetized by state.